The Financing of
Technological Change

Research in Business Economics and Public Policy, No. 10

Fred Bateman, Series Editor

Chairman and Professor
Business Economics and Public Policy
Indiana University

Other Titles in This Series

The Financing of Technological Change

by
Lorne Switzer

UMI RESEARCH PRESS
Ann Arbor, Michigan

338.06
S979

Produced and distributed by
UMI Research Press
an imprint of
University Microfilms International
A Xerox Information Resources Company
Ann Arbor, Michigan 48106

Library of Congress Cataloging in Publication Data

Switzer, Lorne, 1954-
The financing of technological change.

(Research in business economics and public policy ;
no. 10)
A revision of thesis (Ph.D.)—University of
Pennsylvania, 1982.
Bibliography: p.
Includes index.
1. Technological innovations—finance. 2. Research.
Industrial—Finance.   3. Federal aid to research.
I. Title. II. Series.
HC79.T4S94 1985   338'.06   85-16419
ISBN 0-8357-1689-9 (alk. paper)

To my parents

Contents

List of Tables

Acknowledgments

First and foremost, I would like to thank Professor Edwin Mansfield for his tireless encouragement and support of these studies. I would also like to thank Professors Richard Easterlin, Almarin Phillips, Pablo Spiller, Oliver Williamson and Anthony Romeo, as well as Omar Ashur, for their contributions at various stages of the research. Thanks are also due to numerous executives for their cooperation in providing data.

I would especially like to thank my wife Marsha for her devotion and enthusiastic support, and my wonderful children Sarede and Andrew for the inspiration they provided.

I am grateful for the financial assistance I received from Professor Mansfield through a grant to him by the National Science Foundation.

Finally, I would like to thank the journals *Management Science, Research Policy, R & D Management,* and the *Review of Economics and Statistics* for permission to include material that first appeared in their pages.

Introduction

Technological change has played an important role in the growth of industrialized countries in the past century. Financing technological change is an important challenge that faces most nations of the world. In a mixed economy, the resources available for undertaking investments to promote technological change are not present in unlimited supplies at zero cost. Both governments and firms often face severe budgetary restrictions which limit their spending in particular areas. This challenge has two aspects.

First, firms may face restrictions on the resources they devote to R & D and innovation given their internal cash flow limitations and the potential competing uses of funds, as well as capital market restrictions. Government policy may play a role in alleviating this constraint through direct subsidization of firms or through tax credit schemes. The effects of such policies to firms will depend in part on the complex interaction between the sources and uses of funds within the firm as well as the additional constraints on firm behavior imposed by the government.

The second and related aspect of this financial challenge is the determination of the absolute level of investments in new technology required by the firm in order to meet the competition at home and abroad. Budgeting for a given level of commitment to innovation in any given year will be dependent on the costs of innovation, as well as the rate of change over time of these costs. The use of appropriate cost deflators is essential for sound budgeting.

To a great extent these issues have been neglected in the extant literature. However, policy makers have become increasingly cognizant of them. One of the first pieces of legislation enacted by the Reagan administration in its first term was to provide companies with tax credits for increased R & D spending. To the extent that private R & D expenditures are responsive to cash flow effects, such legislation may have its desired effects. The cost of such legislation, of course, is the addition to the mounting deficit caused by forgone tax revenues.

An alternative approach to fostering technological change is direct government support of private R & D. Proponents of this course of action would argue that the forces of complementarity between public and private R & D spending overwhelm any tendencies towards substitutability. Also, they might argue that the costs of this measure in terms of its additions to the deficit be more easy to quantify relative to the tax credit approach. Policy makers in the Canadian government have increasingly favored this approach, though it has lost adherents in the U.S.

It is clear that in dealing with the problems of financing innovation, policy makers within firms and within the government should be guided by realistic models of firm behavior as well as adequate data. This book represents an attempt to provide both in examining the relevant issues.

Our agenda will be as follows. In chapters 1 and 2, we will examine the problem of financial restrictions facing firms in undertaking investments in new technology, and the effects and costs of the government stimuli of direct expenditures and tax credits.

In chapter 1, a general flow-of-funds model explaining aggregate R & D investment behavior along the lines established initially by Dhrymes and Kurz (1967) is developed and estimated using company level data in the U.S. The approach generalizes a number of previous models by allowing for simultaneous interaction between alternative uses of the firm's investment funds and the sources and uses of investment funds. Several conclusions will emerge relating to:

(a) the importance of internal funds relative to external financing for R & D;
(b) the opportunity cost of R & D expenditures relative to capital outlays;
(c) the crowding out of public expenditures by private expenditures; and
(d) the potential impact of direct tax credits including the relationship between the revenue loss to the government and the R & D stimulation induced—the analysis here will be supplemented with references to additional evidence we have obtained in a separate study relating to Canada (a country which has had a fairly long history of tax credits for R & D that are quite similar in nature to the recent U.S. legislation).

The analyses of chapter 1 primarily use firm data and it might be remarked that in examining topics such as the implications of government financing on particular areas of technology such as energy or health, a higher degree of disaggregation is required. Thus in chapter 2 we will proceed to focus on the issue of the effects of government financing of R & D in the important area of energy, using project level data that we have obtained in a field study.

Since the early 1970s, government support of energy R & D conducted in the private sector has been rising at a rapid rate—in fact at a rate that has exceeded all other government R & D budget functions. The rationale for the increased level of government support is that the private sector acting alone will underinvest in energy R & D owing to risk aversion and/or limitations on appropriability of returns. Also, since 1973, the motive of national security has been at the forefront of government decisions regarding energy R & D. Government policy has ostensibly emphasized the support of projects that are long term and high risk ventures, and that are more likely to complement, rather than substitute for, private R & D efforts. To this date, only anecdotal evidence has been forthcoming to allow one to assess whether government funded energy R & D projects conducted in the private sector actually stimulate further private R & D outlays, or instead serve as a retardant to private R & D. Based on the analyses conducted for our sample, we hope to shed some light on this issue in chapter 2.

In chapter 3 we will shift our attention to the problem of the changing costs of investments in innovation in the recent past. At present, none of the cost deflators available to firms or the government is entirely adequate. The Battelle Memorial Institute publishes an annual "cost of research index" based on the methodology developed by H. Milton (1966, 1972) that for reasons to be discussed in this chapter, has severe drawbacks. The National Science Foundation in the U.S. uses the GNP deflator, a measure that is also potentially unreliable. It is essential that more adequate deflators be developed. We will thus proceed in this chapter to construct a number of cost deflators for both industrial R & D as well as industrial innovation, using project level as well as company level data. The data have been amassed from companies that accounted for about one-ninth of all company funded R & D in the U.S.; one of the main objectives of the analysis is to evaluate the nature and extent of biases that ensue when one chooses to rely on the usual proxies for costing R & D. Another objective is to demonstrate a fairly tractable approach which managers could use to infer the rate of increase in relative R & D costs over time using actual expenditure data.

The study then concludes with a review of the main findings and presents some suggestions for further work.

1

Investing in R & D: The Problem of Alternative Sources and Alternative Uses of Funds

Introduction

In this chapter, the firm's R & D budgeting decision process is cast within a simultaneous equation flow-of-funds framework. When a firm invests in R & D it frequently has to balance the potential returns against their costs, in terms of competing uses of funds. In addition, the firm's investments must be financed somehow, either externally through capital markets or through internally generated funds. These constraints have largely been ignored in previous work on the R & D investment decision process. In this study they are made explicit.

The approach here contrasts with most previous work that follows the presumptions of standard financial theory appearing in introductory textbooks. In the standard approach, investment decisions (including investments in technological change) may be dichotomized from financial considerations and investment decisions can be made sequentially, without regard to alternative uses of funds. For the simple textbook approach to be valid, some very restrictive and untenable assumptions must be imposed, such as an absence of market imperfections (which would include investor trading costs, limitations on personal borrowing, personal tax biases, informational costs, flotation costs, agency costs, asset indivisibilities, and limited markets) and bankruptcy costs. Theoretically, it has been demonstrated that once the former are accounted for, clear interdependencies between financing and investment as well as between alternative forms of investment appear.[1] Once bankruptcy costs are recognized, debt management becomes a salient area of concern for corporate managers.[2]

Schematically, the general approach taken here is shown below in figure 1. Unlike much previous work the approach endogenizes or links simultaneously five key decisions/constraints within the firm: the capital expenditure decision (investment in plant and equipment), the R & D investment decision, the dividend payout decision, the external financing decision (constraint) and the internal financing constraint. Previous work has often looked at the R & D decision in isolation

Figure 1. A Schematic Representation of the Model[1]

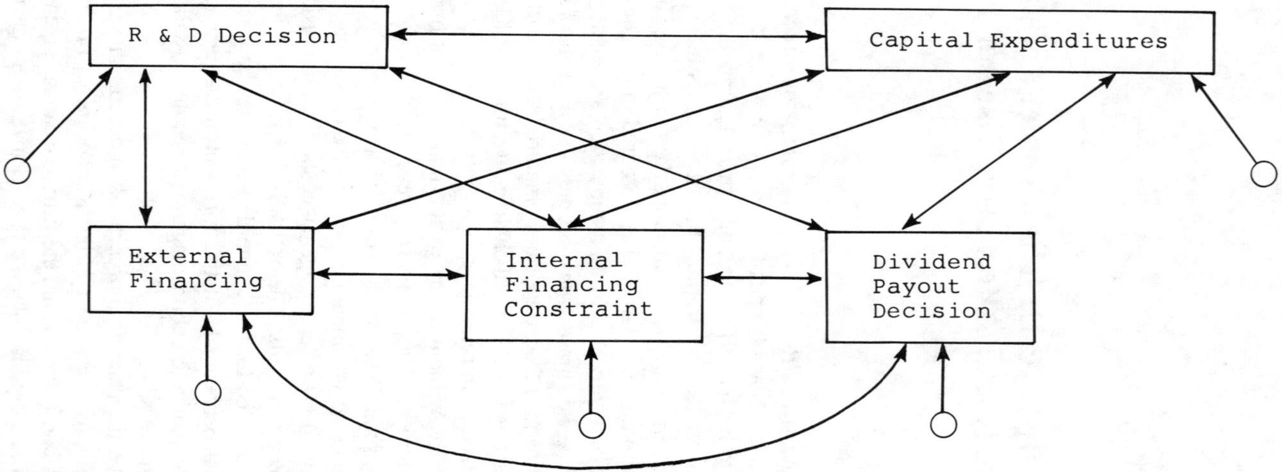

[1] Rectangles denote endogenous variables (determined within the system) while circles represent exogenous variables.

or has failed to account for competing uses of firm's R & D investment and/or the external financing option (constraint).

The Model

Description of the System

The structural schemata chosen might be viewed as a synthesis of the financial flow approaches to modeling capital investment which to date have ignored the simultaneity of the R & D and capital expenditure decisions (for example, Dhrymes and Kurz (1967), and McCabe (1979)) with the simultaneous equation models of investment in R & D that have ignored the simultaneity of investment and financing decisions (Mueller (1967), and Grabowski and Mueller (1972)).

Briefly, the system endogenizes the firm's R & D expenditures, its capital expenditures, its dividend payments, new debt financing and internal financing, and consists of the following equations:

$$(1) \quad RD_i = h_1 (I_i, DIV_i, ND_i, IF_i, CR_4, CRT_i, G_i, DST_i, RDY_i)$$
$$(2) \quad I_i = h_2 (RD_i, DIV_i, ND_i, CF_i, DS_i, DS5_i, INT_i, K_i, IY_i)$$
$$(3) \quad DIV_i = h_3 (RD_i, I_i, ND_i, PRTS_i, DIVY_i, GRT_i, VAR_i, NY_i)$$
$$(4) \quad ND_i = h_4 (RD_i, I_i, DIV_i, CFT_i, RISK_i, PRA_i, BETA_i).$$

With the added identity (I) $IF_i \equiv RD_i + PRB_i + DEP_i$, where the endogenous variables are

RD_i = company funded R & D expenditures[3] by firm i;
I_i = capital expenditures for firm i;
DIV_i = dividend payments of firm i;
ND_i = new long-term debt financing of firm i;
IF_i = internal financing of firm i;

and the exogenous variables are

PRB_i = profits before taxes of firm i;
DEP_i = depreciation and depletion allowances of firm i;
CR_{4i} = average of the four-digit four-firm concentration ratios of firm i's industries of classification;
CRT_i = $(CR_{4i})^2$;
G_i = government R & D awards to firm i;
DST_i = ten-year change in sales (1967–77) for firm i;
RDY_i = previous years R & D expenditures of firm i;

CF_i = after-tax profits plus depreciation and depletion allowances for firm i (lagged 1 year);

DS_i = one-year change in sales of firm i;

$DS5_i$ = five-year change in sales (1972–77) of firm i;

INT_i = average interest rate of firm i (total interest payments to total long-term debt outstanding);

K_i = lagged capital stock of firm i;

IY_i = lagged investment of firm i;

$PRTS_i$ = current profits after taxes of firm i;

$DIVY_i$ = lagged dividends of firm i;

GRT_i = long-term growth in firm i's earnings per share (measured over nine years 1968–77);

VAR_i = standard deviation of firm i's earnings to net worth (measured over a five-year period 1972–77);

NY_i = a dummy variable representing the exchange listing of the company's common stock (equals one if the listing is NYSE and zero otherwise);

CFT_i = after tax profits plus depreciation and depletion allowances for firm i (unlagged);

$RISK_i$ = coefficient of variation of firm i's after tax profits plus depreciation and depletion allowances (measured over the five-year period 1972–77);

PRA_i = the ratio of lagged profits after taxes to total assets for firm i; and

$BETA_i$ = firm i's systematic risk.

This system can be justified in two ways. First, one may argue, as did Dhrymes and Kurz, that "technological and marketing constraints are exogenous to the system and predate the decision process we wish to study."[4] Once we admit the cash flow constraint into such a system, McCabe notes we get the implication that "sources of funds would affect uses of funds positively, and other sources negatively, and conversely."[5]

Secondly, and perhaps more insightfully, (1)–(4) can be derived as equilibrium conditions explicitly from a system of first order conditions for utility maximization in the spirit of Grabowski and Mueller.[6]

Selection of Exogenous Variables

With regard to the choice of exogenous variables to identify the equations of the system, we have in general selected regressors which have explained with some success the behavior of the endogenous variables here in, for the most part, single equation formulations. Thus, the contribution of the approach lies in its examination of the effects of the variables in the context of a simultaneous equation formulation.

The R & D equation. Previous studies of the determinants of R & D investment have emphasized variables which have served as proxies for rate of return and appropriability characteristics. Cash flow, denoted as the sum of after-tax profits plus depreciation and depletion allowances lagged one period, is frequently used to account for possible aversion to external financing of R & D and to serve as an expected returns proxy in line with the reasoning used in support of tax concessions for R & D. However, such treatment ignores the contemporaneous nature of the cash flow constraint to the firm. To capture this in the approach here, internal financing (IF) is specifically embodied as an endogenous constraint, and is expected to have a positive effect on firm R & D investments.

Four-firm concentration CR4 was considered as a means of testing the neo-Schumpeterian approach, though with some reservations. The neo-Schumpeterian position is perhaps most vehemently propounded by J. K. Galbraith, and asserts that large size and structural monopoly provide the most favorable atmosphere for technological progress. It is a variant of Schumpeter's view that transient monopoly, as opposed to competition, is required for progress.[7]

The squared value of this term was also used to test a "diminishing returns" variant of the neo-Schumpeterian approach. Given the somewhat ambiguous evidence of the effects of concentration on R & D in single equation formulations (for example Comanor, Link) as well as in Levin's simultaneous model, no prior prediction on the sign of this coefficient was made.

Next, government support of R & D was included to test the complementarity vs. substitutability arguments that have been raised for government R & D. Given the recent Levin (1980) and Link (1982) findings for fairly recent periods, some degree of complementarity was expected a priori. Inferring the effects of government expenditures on private R & D on the basis of regression analysis using aggregative data is a highly tentative procedure, though. In the next chapter, we will address its shortcomings in more detail and will provide alternative evidence regarding the effects of government expenditures using project level data for energy R & D.

The firm's ten-year change in sales (1967–77) was included to capture uncertainty and demand pull effects recognizing the alternative potential interpretations of this variable.[8]

Finally, lagged R & D was included as a predetermined variable to account for partial adjustment effects, in the spirit of Mansfield's model (1964). Mansfield's explanation for the lagged adjustment process implied by the use of this variable is as follows.

First, it takes time to hire people and build laboratories. Second, there are often substantial costs in expanding (R & D expenditures) too rapidly because it is difficult to assimilate large percentage increases in R & D staff. . . . Third, the firm may be uncertain as to how long expenditures of (desired R & D levels) can be maintained. It does not want to begin projects that will soon have to be interrupted.[9]

The capital expenditures equation. As the main focus of this chapter is concerned with the determinants of nominal R & D investment, an exhaustive review of previous work on the capital investment equation will not be provided. Suffice it to note that Jorgenson relates that the flexible accelerator mechanism has been the "point of departure of the large body of empirical research on investment behavior."[10] In this approach, he notes "changes in desired capital are transformed into actual investment expenditures by a geometric lag function . . . "[11] Hence, if we were to focus on gross investment using this schemata, assuming a constant rate of depreciation we get a functional form such as:

$$I_{G_t} = (1 - \lambda) (K_t^D - K_{t-1}) + K_{t-1}^{[12]}$$

where
$$I_{G_t} \equiv \text{gross investment in year } t$$
$$K_t^D \equiv \text{desired capital stock in year } t$$
$$K_t \equiv \text{actual capital stock in year } t$$
$$(1 - \lambda) \equiv \text{the adjustment coefficient.}$$

The alternative theories of investment that have appeared are thus based on different views on the determinants of K^D, the desired capital stock. Jorgenson recognizes three main approaches:[13] (a) the external financing approach, (b) the internal financing approach, and (c) the output or capacity utilization approach. Supporters of (a) would assert that the optimal investment level is determined by the firm's marginal cost of funds, proxied by some external interest rate. Firms are seen to be quite willing to resort to external financing (when it is available) when pressures on capacity appear. Proponents of (b) would argue that desired capital stocks should be affected by internal funding since at the point where internal sources of funding are depleted (the usual state of the world), the marginal cost of capital schedule becomes highly inelastic. On the other hand, supporters of the capacity utilization approach assert that increased output (proxied by increased sales) are the primary cause of net investment, with financial constraints of no importance. Given the somewhat mixed evidence on the performance of the various theories, all of them will be given some consideration in what follows.

Thus, to account for the capacity utilization approach, one-year and five-year changes in the firm's sales are included as regressors. Next, the cash flow term used in the R & D equation is also incorporated into the capital expenditures equation to account for the internal funds approach. In addition, the firm's average interest rate (measured by the firm's total interest payments divided by long term debt outstanding) is included as a proxy for the firm's relevant interest rate, to account for the external financing approach (as a supplement to the endogenous variable, new debt financing). As with Dhrymes-Kurz and McCabe, its inclusion was due to the absence of an alternative, more accurate, measure of the phenomenon of interest.

Since the equation measures gross investment, the lagged value of the firm's capital stock was used to deal with the effects of replacement demand. As Mueller notes (1967, 66), this variable will "act as an index of the capital intensiveness of the firm relative to other firms in the sample and not as a measure of its present level of capital relative to some ideal quantity [in reference to the Griliches time series approach]." Hence, it is expected that its coefficient will be positive, as was found for example in McCabe's cross sections (1979, 130).

Additionally, lagged investment was also included to account for partial adjustment effects, as discussed in reference to the R & D equation.

The dividend equation. Perhaps the most important work concerning dividend policy that has appeared in the literature is that of J. Lintner (1956). Lintner specified an empirical model in which firms are seen to set dividends in terms of a long-run payout ratio and speed of adjustment[14]; the predominant element which affects current changes in dividends, given the targeted payout ratio, is simply the firm's net earnings after taxes. The model is written as:

(L1) $\qquad D_{it} = a_i + c_i(D_{it}^* - D_{i,t-1}) + U_{it},$

where $\qquad D_{it} \equiv$ actual dividend payments of firm i in period t
$\qquad D_{it}^* \equiv$ desired dividend payments of firm i in period t
$\qquad c_i \equiv$ speed of adjustment coefficient
$\qquad U_{it} \equiv$ random error term

Desired dividend payments are written as:

(L2) $\qquad D_{it}^* = r_i P_{it},$

where $\qquad r_i \equiv$ target payout ratio of firm i,
$\qquad P_{it} \equiv$ current profits after taxes for firm i.

Substituting (L2) into (L1) we get the equation:

(L3) $\qquad D_{it} = a_i + b_i P_{it} + d_i D_{i,t-1} + U_{it},$

with $\qquad b_i = c_i r_i$
$\qquad d_i = (1 - c_i),$

which has been used quite successfully to explain dividend behavior in the past. In fact, some authors have claimed that it "stands among the more thoroughly founded hypotheses in the area of business behavior."[15]

As a further test of this model, current profits after taxes and lagged dividends were adopted as regressors in the dividend equation. Two additional variables used by Grabowski and Mueller (1972) to account for managerial discretionary behavior were also incorporated into the model. The first of these is the firm's long-run growth in earnings per share (measured over a nine-year period);[16] this variable might be expected to lower the level of dividend payments to the extent that growth reflects favorable opportunities for the use of retained earnings (allowing the firm to bypass capital markets for investment purposes). Second, the standard deviation of the firm's ratio of earnings to net worth (measured over a five-year period, 1973–77) was used to account for the presumption that managers attempt to counterbalance the dangers to their security posed by high earnings variability by increasing dividend payouts. In the presence of such behavior, then, this variable can be expected to have a positive coefficient in the model.

The final variable considered in this equation is the exchange listing of the company's stock to account for the claim that NYSE stocks tend to have higher dividend payouts than do AMEX or OTC stocks.[17]

The new debt equation. The new debt equation has received little attention in the literature. A traditional view of the firm's new debt management policy is that an optional debt level exists which balances the tax gains from increasing leverage against the costs associated with the increased probability of bankruptcy as the firm's fixed costs rise.[18] As was mentioned earlier, we have used the firm's BETA as one measure of riskiness and which may be expected to have a negative effect on debt financing. Also, following McCabe, the firm's coefficient of variation of after tax profits plus depreciation was used as a supplementary risk measure.

The firm's cash flow, measured by its level of current after tax profits plus depreciation and depletion allowances was used as in Dhrymes-Kurz and McCabe to capture the firm's need for preserving solvency, and was expected to have a negative sign. Finally, the ratio of the firm's profits to total assets lagged one period was used to capture capital market rationing effects. The notion is that firms with higher rates of return on their assets may have easier access to external financing; hence the expected sign of this variable is positive.

Description of the Data

The data used to estimate the system were derived for a cross section of 125 manufacturing firms in 1977.[19] The industries chosen are chemicals, petroleum, electronics and aerospace. In terms of the selection of individual firms, we have attempted to provide fairly thorough coverage of the various industries, trying to avoid capturing only the largest firms. Completeness of the data was the ultimate criterion for inclusion in the sample; nevertheless, by including industries that had quite variable research intensities (both high and low), a sample that possesses some of the characteristics of manufacturing industry as a whole has been obtained.[20]

Estimation

The model given by linearized versions of the functions h_1, h_2, h_3, and h_4 in our description of the system was estimated for the pooled sample of 125 firms for 1977 using three techniques—ordinary least squares (OLS), two-stage least squares (2SLS), and three-stage least squares (3SLS); the latter approach which accounts for identity (I) as well in the estimation, given our theory, is the consistent procedure, yielding estimates that are asymptotically equivalent to full information maximum likelihood estimates. Dummy variables[21] were included in each equation not captured in the included regressors. As in previous studies of this sort, all of the endogenous variables, as well as the size-related exogenous and predetermined variables were deflated[22] by firm sales to account for the potential for heteroscedasticity and the problem of "extreme values." Since all of the equations of the system are overidentified, and the covariance matrix of the residuals was found to be nondiagonal, it may be most appropriate to discuss the model in terms of the 3SLS estimates in Tables 1.1 and 1.2.

Estimates of the R & D equation. First, looking at the R & D equation, as with a number of previous models (e.g., Link (1982), Grabowski (1968), Grabowski and Mueller), internal financing seems to be one of the central determinants of R & D expenditures.[23] The importance of internal financing is further highlighted by the lack of significance of the coefficient of new debt, which seems to support the contention that external financing of R & D is unimportant due to, e.g., disclosure requirements and the absence of collateral value for R & D outlays.

The evidence here provides indirect support for potential beneficial effects of tax credits embodied in the Economic Recovery Tax Act on R & D expenditures. For to the extent that tax credits enhance firms' internal financing capabilities, they may have their desired effects of stimulating private R & D.[24]

As with the Levin (1980) and Link (1982) studies, neither the concentration varible nor its square appeared significant, which seems to undermine a variant of the simple neo-Schumpeterian theory.

The ten-year change in sales variable provides support for the expectations theories of Mueller (1967) and others. Lagged R & D is also found to be significant, lending some support for a partial adjustment mechanism for R & D, though this result has to be interpreted with caution.[25] The coefficient for capital expenditures is similar to the Mueller and Grabowski-Mueller estimates. Also, the lack of significance of dividends is consistent with Grabowski-Mueller as well as

Table 1.1.   Three-Stage Least Squares Estimates of the
R & D and Capital Expenditures Equations

R & D Equation			Capital Expenditures Equation		
Independent Variable	Estimated Coefficient	Absolute t-Ratio	Independent Variable	Estimated Coefficient	Absolute t-Ratio
I	−0.003	1.302[d]	RD	0.057	0.416
DIV	0.119	0.959	DIV	0.089	0.146
ND	0.004	0.206	ND	0.126	0.988
IF	0.025	1.453[d]	CF	0.071	0.410
$CR4$	−0.016	0.606	DS	−0.107	2.563[a]
CRT	0.012	0.396	$DS5$	0.023	0.682
G	0.084	0.922	K	0.058	1.812[c]
DST	0.009	1.886[c]	IY	0.501	3.253[a]
RDY	0.707	20.375[a]	INT	−0.015	0.717
$D1$	−0.007	2.759[a]	$D1$	0.011	0.715
$D2$	0.003	1.135	$D2$	−0.005	0.408
$D3$	0.004	1.000	$D3$	−0.005	0.303
Constant	0.001	0.172	Constant	0.009	0.390
GOF	0.962		GOF	0.635	

Note: The goodness of fit measure (GOF) is the square of the correlation coefficient between actual and fitted values (Haessel (1978)).
[a]Statistically significant at the 0.01 level.
[b]Statistically significant at the 0.05 level.
[c]Statistically significant at the 0.10 level.
[d]Statistically significant at the 0.20 level.

Mueller's results. Next, from the sign and significance of the coefficient of D_1, it is apparent that firms in the petroleum industry tend to have lower R & D to sales ratios than do firms in other industries.

Finally, consistent with the recent results of Link, Levin, Terleckyj and Levy (1982) and Mansfield and Switzer (1984), no evidence of complete crowding out of private expenditures by government expenditures could be observed. The coefficient of government expenditures exceeds minus one by a highly significant margin, which implies that social R & D expenditures (i.e., public plus private expenditures) rise for an increase in government outlays, pari passu. This result is quite robust with respect to the specification of the exogenous variables in the model, as well as to the method of estimation.

Estimates of the capital expenditures equation. Looking now at the capital expenditures equation, it is apparent that our results are quite consistent with earlier models.[26] Although the cash flow term usually had a positive coefficient in the various specifications tried, it was never significant. The accelerator or capacity utilization hypothesis for capital expenditures seems suspect, given the negative and significant coefficient for the one year change in sales variable, a result consistent with McCabe (1979). Both the sign and significance of the new debt term

Table 1.2.   Three-Stage Least Squares Estimates of the
Dividend and New Debt Equations

Dividend Equation			New Debt Equation		
Independent Variable	Estimated Coefficient	Absolute *t*-Ratio	Independent Variable	Estimated Coefficient	Absolute *t*-Ratio
RD	0.001	0.240	*RD*	0.252	2.254b
I	−0.018	3.607[a]	*I*	0.457	4.140[a]
ND	0.018	2.373[b]	*DIV*	−0.785	1.171
PRTS	0.018	2.644[a]	*CFT*	−0.218	1.406[d]
DIVY	1.042	31.163[a]	*PRA*	0.613	5.749[a]
GRT	0.0002	0.399	*RISK*	0.002	0.635
VAR	0.0005	1.058	*BETA*	−0.031	2.836[a]
NY	0.0005	1.019	D1	0.002	0.144
D1	−0.001	2.430[b]	D2	0.015	1.242
D2	−0.001	2.436[b]	D3	−0.017	1.012
D3	−0.002	1.836[c]	Constant	−0.011	0.540
Constant	0.003	3.720[a]	*GOF*	0.415	
GOF	0.944				

[a]Statistically significant at the 0.01 level.
[b]Statistically significant at the 0.05 level.
[c]Statistically significant at the 0.10 level.
[d]Statistically significant at the 0.20 level.

are consistent with the conclusions of Dhrymes-Kurz and McCabe. Additional weak support for an external financing hypothesis for capital investment is given by the negative coefficient for the interest rate variable, though this variable is difficult to interpret with confidence and, like McCabe's result, is not significant.

The firm's lagged capital stock variable appears to be capturing the effects of replacement demand, as expected. From the lagged investment variable, it appears that actual investment responds to changes in desired levels only gradually, though the speed of adjustment is much more rapid than for R & D. Finally, dividend outlays do not seem to be significant determinants of capital expenditures. Overall the absence of significance of financing variables (*ND, DIV*) contrasts with Dhrymes-Kurz's and McCabe's general results, though not with Fama (1974), and appears to provide some support for the Modigliani-Miller perfect capital markets model.

Estimates of the dividend equation. Proceeding now to the dividend equation in Table 1.2, it is quite clear that Lintner's claim (1956, 106) that "investment requirements as such (have) little direct effect in modifying the pattern of dividend behavior" is not validated here. Lintner's approach may be brought into question by the fact that the reaction coefficient for dividends is not significantly different from zero (since the coefficient of lagged dividends is not significantly different from 1).

New debt and current profits had a positive and significant effect as expected. Long-term earnings prospects, to the extent that they are reflected in growth in earnings per share appear to have no influence on dividend outlays, which contrasts with the Grabowski-Mueller result. The earnings variability term has a positive sign, as would be predicted by the managerial discretion hypothesis of Grabowski-Mueller, though, unlike them, we found that it lacked significance. No strong evidence could be found in support of the claim that stocks listed on the NYSE are more "dividend intensive" than others for 1977. Finally, as in Dhrymes and Kurz, the industry dummy variables appear significant, and thus we might assert, as they do (1967, 458), that "it would not appear proper to deal with this (dividend) relation in simple aggregation terms [à la Lintner]. At least this aspect of inhomogeneity must be taken account of. . . ."

Estimates of the new debt equation. With respect to the new debt equation, current cash flow had a negative effect (as expected), indicating perhaps the solvency preserving function of new debt. Current capital expenditures were positive and significant determinants of new debt issues, which is consistent with McCabe and Dhrymes-Kurz. Somewhat surprisingly (though not from the flow of funds perspective), the coefficient of R & D was also positive and significant. What these results seem to imply is that new debt essentially serves an accommodative role. Once firms decide on their R & D and investment commitments, outside capital will be sought to the extent that it is available. However, the initial commitment to invest does seem to be related to the means of financing. In particular, firms apparently prefer to finance R & D internally.

The profit rate term behaved as expected, providing further support for the Dhrymes-Kurz hypothesis (1967, 462) that there exists "noninterest credit rationing" where more profitable firms have easier access to credit markets. The firm's systematic risk, measured by the *BETA* term, also behaved exactly in accordance with our expectations. The supplementary risk variable serves essentially no meaningful purpose in the model, as was also found by McCabe. The dividend term was unexpectedly negative, as was found in the Dhrymes-Kurz model, as well as in McCabe, perhaps owing to the absence of lasting tangible value of dividends, unlike capital expenditures. Another possible interpretation, adduced by Dhrymes-Kurz (1967, 462), is simply "the fact that stock flotation is an alternative to bond flotation."[27]

Summary and Conclusions

In this chapter we have attempted to model for the first time, using U.S. data, the determinants of R & D investment within the context of a flow of funds, simultaneous equation framework. This approach generalizes a number of previous models of the R & D investment decision process that have appeared in the literature by allowing for simultaneous interaction between alternative uses of R & D funds (dividends and capital expenditures) and between the sources and uses of funds.

Four key observations might be noted. First, although previous models have often recognized the importance of internal funds for supporting R & D programs, and have hinted at the unimportance of external financing for R & D, this is the first model to appear that has perhaps quantified both phenomena. Second, although some previous studies using American data have concluded that R & D expenditures may lower the marginal returns on capital outlays, our results show no such effects. Also, consistent with recent evidence of other authors, no evidence of crowding out of private expenditures by government expenditures could be found. This result is fairly robust to alternative specifications of the exogenous variables of the model, as well as to alternative methods of estimation. Finally, in contrast with Dhrymes-Kurz and others, but consistent with Fama (1974), the Modigliani-Miller result of independence of capital investment from financing could not be rejected in the simultaneous procedures, as opposed to the single equation results.

2

The Costs of Government Support of R & D: The Case of Energy

Introduction

The problem of measuring the costs and effects of government funding on private research and development has not been addressed to any significant extent in the industrial organization literature on the economics of technological change in the past twenty years. This is unfortunate. As M. Kamien and N. Schwartz remarked in a famous survey paper, "since antitrust, patent and copyright law and government financing (all) may influence the course and rate of technological advance, determining how they do so is of interest to policy makers as well as theorists."[1]

The rationale for government funding of R & D in industry usually proceeds along the following lines. First of all, many research activities have pure public good attributes associated with them. For example, as E. Mansfield remarks, "collective consumption activities such as national defense and the space program involve the Federal government as the sole or principal purchaser"[2] of the final goods and services. Since the Federal government has the primary responsibility in these activities, he notes "it must also take primary responsibility for the promotion of technological advance in the relevant areas."[3]

Second, the Federal government may intervene in areas where the private sector would underinvest in R & D for goods normally purchased in the private sector. As the *Economic Report of the President* of 1972 relates, R & D often entails nonappropriable rents, especially for basic research. The report claims, "although an investment in R & D may produce benefits exceeding its costs from the viewpoint of society as a whole, a firm considering the investment may not be able to translate enough of these benefits into profits on its own products to justify the investment."[4] Mansfield clarifies this point further. "The results of R & D can be appropriated only to a limited extent because of the riskiness and costliness of R & D. . . . Industrial firms will invest [in basic research] less than is socially optimal since the results are unpredictable and usually of little direct value to the firm supporting the research although potentially of great value to society."[5,6]

In the past, the Federal government has funded well over one-half of all industrial R & D performed in the U.S., although in recent years the proportion of total R & D funding accounted by the Federal government has dwindled somewhat.

It would seem to be an important task to examine the implications (both in terms of production and transaction costs) of the Federal government's support of R & D. In an important survey, Zvi Griliches (1979) posed the question, "Does Federal R & D substitute for or complement private R & D investment?" To date, there have appeared a number of analyses of the interaction of Federal and corporate funding of research and development which might be brought to bear on this issue.

An early attempt to measure the effects of government funding on private R & D is that of Blank and Stigler (1957). Their analysis stems from their consideration of factors which explain the increased proportion of the U.S. labor force accounted for by engineers and chemists, through the first half of the twentieth century. In looking at the determinants of private R & D efforts (a factor affecting the demand for scientist-engineers) they postulated a range of possible effects of the enhanced R & D activity of the Federal government.

For the pure complements case, it was suggested that "private businesses first take on government research contracts, as a result become persuaded of the benefits of research, and then embark on private research also—so that the government contracts serve a sort of pump-priming function."[7] For the pure substitution case, they envisioned a scenario in which "research that the businesses had been conducting might be simply shifted to public contracts, so these contracts would constitute no net additional demand."[8] On the other hand, an intermediate case of independence of government and private research funds was also deemed plausible on a priori grounds so that research undertaken with government funds "constitutes a net addition to total research."[9]

Blank and Stigler's methodology, which is quite indirect, proceeded as follows. First, they subdivided a large sample of firms into two groups: (a) those that conducted no government research in their labs, and (b) those that did. For the firms in each group, they computed the ratio of total scientists and engineers engaged in private research to total employment; they hypothesized that to the extent that public R & D efforts are substituted for privately funded R & D, the ratio computed for firms in group (a) should be greater than that for firms in group (b). This hypothesis was borne out by all but 2 of 17 industry groups. For the industries taken as a whole, furthermore, they found that the average ratio of scientists and engineers to total employment was 1.4 percent in group (a) firms and 0.7 percent for group (b) firms. Noting further that the ratio of scientists and engineers in group (b) engaged in government R & D averaged 0.8 percent, their logical conclusion was that about one-eighth of all government R & D apparently provides a net addition to society's R & D efforts.[11] Hence the pure substitution hypothesis was ruled out.

Next, in considering only firms that performed both public and private R & D work, they found that industries with high ratios of engineers and scientists in private research to total employment also had high ratios of engineers and scientists engaged in government research to total employment. On stratifying firms according to size, such a relationship was strongest for large firms. This then seemed to imply fairly strong complementarity between public and private R & D. The authors are wary of such an interpretation, but nevertheless conclude that "the crude estimate of substitution derived from the aggregate data grossly exaggerates the forces making for a substitutive relation."[12] Overall then, their analysis appears to deny any net effects of government research on society's overall commitment to R & D. However, the estimates may have been somewhat biased, since firms were excluded from the analyses when their ratios of scientists and engineers to total employment exceeded ten percent.

Since the Blank and Stigler study there have appeared in the literature two other somewhat more direct approaches to analyzing the effects of government expenditures on private R & D. The first approach embraces studies that are essentially "reduced-form" investment approaches, while the second might be deemed the productivity of research approach.

The reduced-form investment approaches, which were discussed in the previous chapter, are primarily single equation multivariate models of the determinants of privately funded R & D expenditures for firms or industries, with government R & D allocations appearing as an explanatory variable (usually treated as exogenous, with the exception of the Levin (1980) paper) of the various models. In these formulations, the degree of substitutability/complementarity of private and public R & D is determined by examining the signs and significance of the coefficients obtained for the government expenditure variables in the empirical implementation of the models. These approaches are termed "reduced-form" since they do not account for the interaction amongst alternative uses of investment funds for firms (or industries), nor do they account explicitly for financing constraints, problems which we attempted to remedy in chapter 1. As we showed, complete substitution or "crowding out" of private R & D expenditures by public expenditures has not been an observed feature of the empirical implementation of these models. Nor was crowding out observed in our analysis of the determinants of R & D investment that extends and generalizes these models.

Studies dealing with the productivity of research approach, on the other hand, attempt to examine the differential effects of privately funded and government funded R & D allocations on the rate of growth of some measure of output or welfare.

One study in this spirit is that of W. A. Leonard (1971). Leonard looked at the relationship between various measures of industry growth rates (sales growth, assets growth, growth of net income, etc.), corporate research intensity, and Federal funding for sixteen industrial groups (which encompassed almost all manufacturing activity) in the U.S. over the period 1957–63. Leonard uses corporate R &

D funds to net sales as a proxy for company research intensity as he feels that the use of manpower variables, which do not account for the high costs of equipment, facilities, and supporting personnel, does not allow one to adequately capture development efforts that are supported by the firm.[13] In his analysis, he finds a strong positive correlation between industry growth rates (however measured) and corporate research intensity in all the industries. A significantly positive relationship between Federal R & D intensity and growth rates appears when one excludes two industries from the analysis: aircraft and missiles, and electrical equipment. In these two industries, government funded R & D apparently detracts from the productivity enhancing effects of private R & D.

One possible explanation Leonard provides for his results is that it may be the case that a disproportionate concentration of Federal R & D support in the aircraft and missiles and electrical equipment industries produced an effect of diminishing returns.[14] At any rate, he gives an alternative explanation for his results: large military and space R & D contracts that are supported by the government tend to engender excessive wastes.[15] In this regard, O. Williamson suggests "inferior research and contracting practices of the military services" as another cause.[16]

The second of the "productivity of research" studies that should be mentioned is that of N. Terleckyj,[17] who extends previous work which found (using industry data for 1948–66) that privately funded R & D was significantly associated with industrial productivity growth, but that government financed R & D was not. Using the Cobb-Douglas specification with disembodied technical change and adjusting his original productivity measure for the use of human capital as well as intermediate goods, his findings were basically unaltered, even after allowing for a few alternative measures of productivity. Total R & D was still a significantly favorable factor affecting productivity, while government R & D considered alone was not. S. Globerman attempts to refute any implication of inefficiency of government R & D suggested by this model as follows:

> To the extent that Federally financed R & D is primarily directed towards improvements in product quality as opposed to cost reduction, the methodology used in deriving industry productivity estimates could contribute to the finding that government financed R & D is not significantly related to productivity change. Furthermore, while the period 1948–66 might be long enough compared to the time lag one would expect between privately financed R & D intensity ratios in 1950 and their productivity effects, it might be too short to fully incorporate the effects of Federally financed R & D which is presumably aimed at effecting greater changes in underlying production conditions . . . Leonard's (1971) study is subject to the same sorts of criticisms.[18]

In a companion paper, Griliches uses 1957–65 corporate level data for 883 manufacturing companies to model, like Terleckyj, the returns to R & D, using once again the Cobb-Douglas disembodied technical change model. From his regressions, he notes he was "unable to discover any direct evidence of the superiority of company-financed R & D as against Federally financed R & D in affecting the growth in productivity."[19] As in the Leonard study, in the electrical equipment and aircraft and missiles industries, government R & D apparently lacked beneficial effects on relative growth rates; in these industries the lowest estimates of "excess returns" to R & D were found. We note that Griliches' explanation for this phenomenon is quite similar to Leonard's, as he states:

> It may well be the case that within any company a dollar is a dollar, irrespective of the source of financing, but that in these two specific industries the externalities created by large Federally financed R & D investments and the constraints on the appropriability of the results of research that may have been associated with such investments have driven down the realized private return to R & D significantly below its level prevailing in other industries.[20]

To briefly summarize, it seems that the investment function analyses of R & D that have appeared (including our own version) are on the whole inconsistent with a complete crowding out interpretation of the effects of government support of R & D. Also there is not much evidence from the productivity of research approaches that in aggregate, government support of R & D adversely affects economic growth. These efforts have their shortcomings, however.

Perhaps the main problem concerns potential aggregation biases. One might assert that a considerable amount of variability may be concealed by them, since they only capture industry and company-wide effects. For the purpose of examining the effects of government expenditures on any given area of R & D such as health or pollution abatement or energy R & D, as opposed to R & D in general, a much more disaggregated data base is clearly necessary. Studies that merely lump together all of the diverse sets of activities conducted by firms (or industries) essentially treat the different activities as perfect substitutes, when in fact they are not. As a means of dealing with this aggregation problem, as well as a means of studying a topic that is quite timely and important in its own right, we will focus solely on the effects of government expenditures on the set of activities embraced by energy-related research and development.[21]

Since the early 1970s, Federal government involvement in energy-related R & D has grown at a rate far surpassing the growth rates of all other budget functions. Since 1977, Federal energy R & D has been almost on a par with health R & D, being surpassed only by national defense and space R & D in terms of total expenditures. In this period, corporate funding of energy-related research and development has risen dramatically as well (as we will see in this chapter). An article in *Business Week* states succinctly: "the energy crunch, in fact, heavily influences the research efforts of just about every major industry in the U.S. and will continue to do so for years to come."[22]

In the following section, we shall attempt to delineate the evolving position of energy R & D in the overall research priorities of the Federal government and private firms. Next, we will attempt to analyze from a microanalytic perspective the interaction of Federal and private energy-related R & D expenditures using confidential data that we have obtained from a rather detailed questionnaire and from various interviews with senior R & D executives from a small sample of firms which includes some of the largest performers of energy-related R & D in the U.S. A major objective of the analysis will be to shed some light on the question of substitution vs. complementarity for government and private R & D as raised by Griliches. Furthermore, we will attempt to examine some of the determinants of the extent of substitutability/complementarity of private for public funds for energy-related R & D.

In the final section of this chapter, we will summarize the results and discuss some of the policy implications that are predicated by the analyses. The chapter then concludes with a discussion of the limitations of the approach.

Government-Industry Interaction in the Performance of Energy-Related R & D Expenditures in the U.S.

Recent government policy in regard to energy R & D has ostensibly tried to complement work done in the private sector by providing direct financial support to long term, high risk R & D programs. Short term and low risk ventures are supposedly encouraged through rising energy prices, tax credits and regulatory incentives.[23] This emphasis has disturbed a number of prominent observers in industry. R. W. Schmitt and P. J. Stewart feel that these Federal priorities will have two undesirable consequences: the fraction of new technical developments actually commercialized will go down, and the risk of commercialization of new energy technologies to individual companies will go up.[24] They feel that the bias to long term research will hamper existing R & D programs which emphasize commercialization. "When new information and concepts are continually being generated," they remark, "[this will create] doubts about the best direction for commercialization . . . when long range funding increases rapidly it tends to preempt shorter range funding."[25] They summarize the argument as follows:

> The result is that the commercialization of energy technologies is often not recognized as something that must be lived with all the time, as is certainly the case in almost all industrial R & D, but rather as something that can be deferred until all the alternatives are explored and the climate is right.[26]

In the next section we will present some evidence on the ability of various Federally financed energy-related R & D projects to complement or preempt work done in the private sector. First it seems necessary to describe the evolving position of energy R & D in the overall R & D priorities of the Federal government and private firms in the recent past.

As we stated at the outset, fostering energy-related R & D has been a central interest of the Federal government over the past several years. Since 1977, Federal government expenditures on energy R & D have been on almost equal footing with health, surpassed only by space research and technology and national defense. As can be seen from Table 2.1, between 1971 and 1979 government energy R & D grew by 522 percent (in current dollars), far surpassing the growth rates of all other government priorities. Over this period, the rate of growth of all government R & D funding was only 87 percent, while the median growth rate for all budget functions was 104 percent.

With respect to the allocation of energy R & D on a more disaggregative level, unfortunately, no comprehensive breakdown of all energy-related R & D is available. However, a number of inferences might be made from the NSF survey of science resources series data (based on the RD-1 questionnaire). We note that corporate funding of energy-related R & D has grown at a rapid pace, rising by 197 percent from 1973 to 1978, compared to the growth rate of all corporate R & D over this period of 69 percent. Furthermore, over this period corporate funding has remained the predominant source of energy-related R & D support, representing 62 percent of all energy-related R & D expenditures in 1973 and 60 percent in 1978, as can be seen from Tables 2.2 and 2.3. The relative importance of government support of energy R & D varies substantially across industries. We note for example that government funded energy R & D represented about 74 percent of all energy R & D in the electrical equipment industry in 1977 and 1978 and only 23 and 19 percent of the respective amounts for petroleum refining; NSF records virtually no government support of energy R & D in chemicals, on the other hand, in these years, as can be seen in Table 2.2.

In addition, from Table 2.2 we see that two industries have commanded the largest shares of government energy R & D funds, electrical equipment and petroleum refining; these industries absorbed at least 63 percent of the government's energy R & D budget to industry in 1973 and approximately 56 percent of the respective amount in 1978. These two industries have also predominated in terms of corporate funding of energy-related R & D; together they accounted for approximately 40 percent of all corporate funds for 1977 and 1978. Overall, these two industries performed 62 percent of industrial energy-related R & D in 1973 and 46 percent in 1978. If we were to add chemicals and allied products to this group the corresponding percentages for energy-related R & D performed would rise to 67 percent in 1973 and 52 percent in 1978.

Disaggregating by type of energy-related R & D performed, we see in Table 2.4 that from 1973 to 1978 a marked change in emphasis occurred. Nuclear energy R & D, which represented 50 percent of all energy R & D performed in industry in 1973, fell to 33 percent of the total in 1978. Similarly, the relative share of fossil fuels R & D fell over this period from 43 percent in 1973 to 28 percent in 1978. On the other hand, other types of energy-related R & D, such as geothermal, solar, conservation and utilization, have risen to preeminence, from a mere 7 percent share of all energy-related R & D in 1973 to a share of 39 percent in 1978.

Next we will examine trends in the concentration of energy-related R & D

Table 2.1. Federal R & D Funding by Budget Function:*
Fiscal Years 1971–81
(Dollars in millions)

| | Actual | | | | | | | | | Estimates | | | | | |
	1971	1972	1973	1974	1975	1976	1977	1978	1979	1980 January Request	1980 March Reduction	1980 Revised	1981 January Request	1981 March Reduction	1981 Revised
National defense	8,110	8,902	9,002	9,016	9,679	10,430	11,864	12,899	13,791	15,002	−43	14,959	18,135	−19	18,117
Space research and technology	3,048	2,932	2,824	2,702	2,764	3,130	3,365	3,481	3,969	4,606	—	4,606	5,119	−201	4,918
Health	1,288	1,547	1,585	2,069	2,170	2,351	2,629	2,968	3,401	3,682	−32	3,650	3,887	−94	3,792
Energy	556	574	630	759	1,363	1,649	2,562	3,134	3,461	3,834	−69	3,765	3,799	−124	3,675
General science	513	625	658	749	813	858	974	1,050	1,119	1,246	—	1,246	1,435	−64	1,371
Natural resources and environment	416	479	554	516	624	683	753	904	1,010	1,090	—	1,090	1,144	−4	1,140
Transportation	728	559	572	694	635	631	709	768	799	871	−12	860	917	−41	876
Agriculture	259	294	308	313	342	383	457	501	552	604	−3	602	640	−6	634
Education, training, employment, and social services	215	235	290	236	239	255	230	345	354	497	−39	457	639	−299	341
Community and regional development	65	66	78	82	93	109	101	92	127	126	−8	118	143	−6	137
International affairs	32	29	28	24	29	42	66	57	117	127	—	127	135	—	135
Veterans benefits and services	63	69	74	85	95	98	107	111	123	126	—	126	135	−5	130
Commerce and housing credit	90	50	50	51	65	69	71	77	92	107	—	107	119	−5	114
Income security	145	106	106	71	72	48	55	67	57	63	—	63	80	—	80
Administration of justice	10	23	33	35	44	35	30	44	47	48	—	48	46	—	46
General government	7	8	7	9	12	12	13	18	23	21	—	21	23	—	23
Total	15,543	16,496	16,800	17,411	19,039	20,780	23,984	26,517	29,040	32,050	−205	31,845	36,397	−868	35,528

*Listed in descending order of 1981 budget authority. Data for 1971–77 are shown in obligations; data for 1978–81 are shown in budget authority.
Note: Detail may not add to totals because of rounding.
Source: National Science Foundation.

Table 2.2.   Federal Funds for Energy Research and Development
by Selected Industry: 1973-78[1]
(Dollars in millions)

Industry	SIC Code	1973	1974	1975	1976	1977	1978
Total		$385	$482	$622	$754	$951	$1,214
Textiles and apparel	22,23	0	0	0	0	(²)	(²)
Paper and allied products	26	(²)	(²)	(²)	(²)	0	0
Drugs and medicines	283	(²)	(²)	(²)	(²)	0	0
Other chemicals	284-85,287-89	(²)	(²)	(²)	(²)	0	0
Petroleum and refining	29	(²)	(²)	(²)	(²)	67	116
Primary metals	33	(²)	(²)	(²)	(²)	15	18
Electrical equipment	36	244	293	349	463	495	569
Radio and TV receiving equipment	365	0	0	0	0	0	0
Other transportation equipment	373-75,379	(²)	(²)	(²)	(²)	0	0
Aircraft and missiles	372,376	(²)	(²)	(²)	(²)	108	201
Scientific and mechanical measuring instruments	381,82	(²)	(²)	(²)	(²)	0	0
Nonmanufacturing industries	07-17,41-67-737 739,807,891	42	52	67	71	115	146

¹ Data not available for 1972.
² Not separately available but included in total.
³ Less than $0.5 million.
Source: National Science Foundation

funds among firms. Unfortunately no direct evidence could be uncovered with regard to the distribution of government funds for energy-related R & D by firm size. One important industry observer remarked that over time, the government has changed its practice of providing large individual contracts to the main firms in industry to a practice of "breaking up their projects into small pieces" and spreading them among several contractors.[27] We note that Federal regulations have required that efforts be devoted to "publicizing research and development procurement actions to attract small business."[28] Furthermore, government policy in the selection of sources for R & D procurement has required since 1970 that "agencies shall exercise all reasonable efforts to increase the number of qualified small business sources to perform research and development contracts and shall encourage the participation of small business sources, as well as other sources in such procurement actions."[29] Nevertheless, as can be seen in Table 2.5, the largest firms maintained a predominant position in the procurement of government contracts as late as 1978. Within the petroleum refining and electrical equipment industries, the concentration of Federal funding for 1978 is quite marked, as shown in Table 2.6. In these two industries, the distribution of corporate R & D funding is also quite concentrated among the top firms, as is shown in Table 2.7. We also

Table 2.3.   Company Funds for Energy Research and Development
by Selected Industry: 1973-78[1]
(Dollars in millions)

Industry	SIC code	1973	1974	1975	1976	1977	1978
Total		$624	$857	$1,152	$1,319	$1,648	$1,855
Textiles and apparel	22,23	0	0	0	1	1	1
Paper and allied products	26	(2)	(2)	(2)	(2)	1	2
Drugs and medicines	283	(2)	(2)	(2)	(2)	(2)	(2)
Other chemicals	284-85,287-89	(2)	(2)	(2)	(2)	(2)	(2)
Petroleum and refining	29	(2)	(2)	(2)	(2)	469	539
Primary metals	33	(2)	(2)	(2)	(2)	37	31
Electrical equipment	36	74	96	115	122	174	198
Radio and TV receiving equipment	365	0	0	0	0	0	0
Other transportation equipment	373-75,379	(2)	(2)	(2)	(2)	(2)	(2)
Aircraft and missiles	372,376	(2)	(2)	(2)	(2)	57	83
Scientific and mechanical measuring instruments	381,82	(2)	(2)	(2)	(2)	9	11
Nonmanufacturing industries	07-17,41-67,737 739,807,891	39	48	31	66	82	101

[1] Data not available for 1972.
[2] Not separately available but included in total.
Source: National Science Foundation

note, however, that the share of company and government R & D performed by the leading eight firms exceeds their share in net sales of the industries.[30] One might infer from these figures that larger firms are more progressive than smaller firms; at the same time one might claim that government funded energy R & D apparently complements private energy R & D, since where government R & D is high in two of the main energy-related R & D performing industries, so is private R & D. However, this ignores the fact that not all the R & D in these industries is energy-related. Indeed in 1973 only 11 percent of the R & D performed in the electrical equipment industry was energy-related, while the corresponding figure for the petroleum industry was 61 percent. To infer complementarity/substitutability/independence between public and private energy-related R & D one thus needs data at a much more disaggregated level, such as will be presented in the next section.

At any rate, a policy of providing R & D contracts almost exclusively to the largest firms in industry has had some precedents in history and may have some justification from a transactions costs perspective. For historical precedents, E. B. Roberts states that for DoD research and development, most contracts have been

Table 2.4.   Industrial Expenditures for Energy Research and Development
by Primary Energy Source: 1972-78
(Dollars in millions)

Primary energy source	1972	1973	1974	1975	1976	1977	1978
Total	$750	$1,009	1,339	$1,774	$2,073	$2,599	$3,069
Fossil fuels	(¹)	438	516	550	605	765	859
Oil	(¹)	297	329	333	381	454	(¹)
Gas	(¹)	51	74	66	68	94	(¹)
Shale	(¹)	12	18	19	24	35	(¹)
Coal	(¹)	49	65	109	127	177	(¹)
Synthetic fossil fuels	(¹)	(¹)	21	50	74	116	(¹)
Mining	(¹)	(¹)	5	9	10	9	(¹)
Other	(¹)	(¹)	39	50	43	52	(¹)
Other fossil fuels	(¹)	29	30	23	5	5	(¹)
Nuclear	(¹)	501	601	700	799	935	1,006
Fission	(¹)	476	567	659	741	852	(¹)
Fusion	(¹)	25	34	41	58	83	(¹)
All other energy	(¹)	70	222	524	669	899	1,204
Geothermal	(¹)	1	2	6	13	24	(¹)
Solar	(¹)	2	7	19	43	65	(¹)
Conservation and utilization	(¹)	(¹)	137	435	528	694	(¹)
All other sources	(¹)	67	76	64	85	116	(¹)

¹ Category not available for reporting purposes.
Source: National Science Foundation

awarded on a sole source basis to large firms.[31] He cites as evidence the Harvard Business School study of weapons acquisitions, wherein it was found that 63 of the 99 principal contract selection actions were made without competition.[32] In addition, he asserts that contracts that often do in fact follow formal competition procedures "have in reality been committed prior to the receipt of company proposals."[33]

The transactions cost reasoning for allocation of government R & D funds to large firms might proceed somewhat as follows. Many government sponsored R & D projects can be regarded as recurrent contracts which entail a high degree of uncertainty and the commitment of semispecific or idiosyncratic investments in human and physical capital. For cost efficiency to prevail, a bilateral relationship between firms and the Federal government (as described by Williamson's schemata for intermediate product markets[34]) may be called for. A transaction specific relationship which precludes competitive bidding may be the only way to ensure the contracts are carried through in the presence of high uncertainty (with the necessary adaptive sequential decision making procedures providing a means to economize on bounded rationality) and limited alternative uses for the

Table 2.5.   Share of Federal Funds for R & D to Industry by Firm Size
1963, 1973, 1978

Size of firm	Percentage of Federal Funding		
(# of Employees)	1963	1973	1978
Less than 1,000	3	4.1	2.2
1,000 to 4,999	6	4.4	3.5
5,000 to 9,999	91	2.2	3.5
10,000 to 24,999		6.3	7.8
25,000 or more		83.0	83.0

Source: National Science Foundation

Table 2.6.   Distribution of Federal R & D Funds in 1978

Industry	Top 4 Firms	Next 4 Firms	Next 12 Firms
Electrical Equipment	65%	14%	15%
Petroleum	93%	4%	3%

Source: National Science Foundation

Table 2.7.   Share of Company R & D Funds and of Net Sales
for the Top Eight Companies in the
Petroleum and Electrical Equipment Industries

Industry	Share of Net Sales (%)	Share of Company Funded R & D (%)
Electrical Equipment	50	71
Petroleum and Refining	66	83

Source: National Science Foundation

human and physical resources. In addition, such a bilateral relationship provides a means of safeguarding against opportunism, by eliminating the hazards posed by "fly by night" R & D contractors.

It is almost unambiguous from the reasoning of transactions cost analysis that for DoD and NASA contracts, where alternative uses to the firms' "idiosyncratic" investments are absent (new atomic weapons are not sold under free market conditions), a strong case for bilateral relationships between major firms and the government for R & D exists. For some energy-related R & D, however, where the investments necessary for a given government contract may be simply duplications of the firm's own efforts, the bilateral governance structure may not be cost efficient.

From a more basic level, one might conjecture that providing government energy-related R & D projects that are technologically and economically connected with projects within the firms' own private R & D portfolios (e.g., in terms of their contribution to the firms' overall technological objectives, their having resources that may be shifted back and forth among private projects at will, etc.) may be a means of supporting substitutability of private and public R & D expenditures, at least with regard to the performance of the projects themselves. A more important prospect, on the other hand, is that such an environment (i.e., where public and private R & D projects are technically and economically interconnected) may yield a superior spawning ground for future projects in which the firm invests its own funds. From a dynamic standpoint, this second prospect seems most worthy of investigation.

Overall, then, resolution of the issue of the relative dynamic effects of a mechanism of allocating government R & D contracts to large firms (that tend to have larger absolute R & D programs[35]) is strictly an empirical matter that cannot be sufficiently examined using aggregative (firm or industry level) data alone. It seems important that in examining the important dynamic issue of the determinants of "spinoff" of private R & D by public R & D, more attention be paid to basic issues concerning the projects themselves (such as the extent to which they are integrated within the firm's overall R & D program). Given the disaggregative nature of the data we have at our disposal, these issues can and will be examined.

In the section that follows, we will devote some effort to modeling the determinants of substitutability/complementarity of privately funded and publicly funded energy-related R & D in the U.S.

The Interaction between Public and Private Energy-Related R & D Expenditures in the U.S.

In this section we will examine the interaction between public and private energy-related R & D expenditures using data which we have collected for a sample of energy R & D projects conducted within the private sector under the sponsorship of the Federal government. Our approach seems superior to previous attempts to model the effects of government expenditures on R & D in at least two respects.

First of all, there are the benefits of disaggregation that we hope to have achieved. For the project level data that we use have provided a means of completely isolating many of the important phenomena of interest. Griliches (1979) advocates the use of micro data at the individual firm or establishment level in studies dealing with the productivity enhancing effects of R & D.[36] However, even firm level data may conceal a great deal of important information and systematic variability that may potentially be of interest. For example, as we noted earlier, were we to focus merely on aggregate R & D expenditures at the firm level, there is the danger that the different types of R & D performed by any given corporation (defense R & D, health R & D, energy R & D) are liable to be treated as homogeneous entities, when in fact they are not. Differences in the nature of R & D performed in any given firm, then, are likely to prevent the analyst who has only aggregative data in his (her) possession from detecting the effects of government stimuli on an individual activity that may be of interest.

The second favorable aspect of our approach is that both ex ante and ex post data are used. As will be shown, we were able to address fairly rich counterfactual statements regarding the firms' desired as well as actual investment behavior for R & D, in response to changes in the availability of government funding. The reduced-form models discussed in chapter 1, which rely solely on ex post data, permit one to examine trends in actual responses, which might be expected to confound unintentional behavior with the behavior that is relevant for public policy purposes.

Let us review now the data which serve as the basis for our analyses.

A Description of the Data

The data which we use were obtained from responses to questionnaires that were completed by senior R & D officials in eleven firms and concerned a total of 41 energy-related R & D projects that were supported by the Federal government and which were performed at various times between 1973 and 1980. Due to the proprietary nature of some of the questions asked, to ensure the broadest scope of participation, a commitment was made to refrain from identifying the firms providing information. The industry groupings which the firms represent include petroleum refining, electrical equipment, and chemicals and allied products. As was shown previously, these three industries have been leading performers of energy-related R & D and together have also been leading spenders of Federal funds for energy-related R & D. Further, we note that our sample of companies (which includes both large and small firms) accounted for approximately 55 percent of all company funded R & D in the petroleum industry in 1976,[37] 23 percent of company funded R & D in the electrical equipment industry in 1976, and less than 5 percent of the respective amount in the chemical industry in 1976.

With regard to the representativeness of the firms, we note that for the petroleum firms, the total company funded R & D to sales ratio of our sample for 1976 is 0.4 percent, which is identical to the ratio for the petroleum industry population as a whole for 1976.[38] Similarly, the company funded R & D to sales ratio for the electrical equipment firms of our sample is 2.4 percent, which is also exactly the same as the population figure for the electrical and electronics industry in 1976. Finally, the company funded R & D to sales ratio of the chemical firms of our sample is 2.4 percent, which is very near to the figure for the chemical industry as a whole of 2.5 percent in 1976. Although our sample is surely not representative of the universe of manufacturing firms, in light of the above it is probably incorrect to claim that the firms are completely unrepresentative of industries that are of major importance in terms of their performance of R & D (energy-related and nonenergy-related), as well as their total economic significance.

The energy sources pertinent to the projects were fairly diverse and included nuclear, oil and gas, coal, shale, and solar technologies. The median duration of the projects was two years and they ranged in total cost from $40,000 to $30 million.

The extent of substitutability/complementarity/independence of government and private energy R & D expenditures demonstrated by the projects was measured from three different perspectives. First we looked at the net incremental impact of the projects on the firms' privately funded R & D efforts. More specifically, we asked for each project: If the firm had not accepted the project, what would have been the effect on the firm's total R & D expenditures? That is what percentage of the amount spent on this project would the firm have spent on R & D (though not necessarily on this particular project) from its own funds if it had not accepted this project?[39] To further clarify, a response in the half-open interval [0,100) would indicate that accepting the project led to some increase in the firm's total R & D allocations, equal to the full amount actually spent on the project by the government in the event of a response of 0. A response of 100, on the other hand, would indicate that accepting the project (with all of the restrictions this might entail, as will be discussed further) led to no net increase in the firm's total R & D allocations at all (i.e., government sponsorship of the project provided no stimulus at all to the firm's total R & D budget).

Second, we looked at the extent to which the firms would have conducted the work specific to the projects themselves, in the absence of government support. Here, officials were asked: "If Federal funding for this project had not been available, what percentage of the work involved in this project would the firm have carried out with its own funds?"

Third, we looked at the extent to which projects "spun-off" ideas that stimulated private R & D investments. In this regard, we asked for each project, "Did the results of this project suggest R & D projects into which the firm invested its own funds?"

R & D executives and analysts have in the past voiced a number of features of government sponsored projects that are thought to impinge on the degree of substitutability/complementarity/independence between public and private R & D that we have been able to measure and test in various models. One variable considered is the expected (ex ante) and ex post technical integration of the project within the firm's R & D portfolio as reflected by the degree to which government sponsored projects contributed to the firms' technological objectives. To measure the extent of ex ante integration of the projects, reflected by their contributions to the firm's technological objectives, relative to privately financed projects, in each case, officials were asked *"When the decision was made to carry out the project,* the firm believed that its contribution to the firm's own technological objectives was about _____ percent of the contribution that would have been made if the same amount had been spent on nonfederally funded energy R & D." To measure the ex post technical integration of projects, reflected by commensurability with the firm's technological objectives relative to privately financed projects, the R & D officials were to respond to the question: *"When the project was completed,* the firm believed that its actual contribution to the firm's own technological objectives was about _____ percent of the contribution that would have been made if the same amount had been spent on nonfederally funded energy R & D."

Responses to the above questions were presumably tempered by the expectations regarding the ability of the firms to appropriate the knowledge acquired during the performance of the projects for their own uses. In performing energy R & D contracts for the Federal government, patent and license rights reside with the government, not the firm. Nevertheless, the personnel working on the projects may derive some new insights from these projects which may be of use to commercial R & D. In addition, firms may request that invention rights may be waived to the firm completely, provided that certain conditions are met.[40]

The next variable we considered that has been mentioned as being related to the substitutability/complementarity issue is the extent of economic integration of the projects within the firm's R & D program reflected by the extent that resources devoted to a government sponsored project were kept separate from projects that were privately financed. We might sensibly conjecture that Federal R & D projects whose resources are segregated from privately sponsored projects would be less likely to complement the firm's private efforts than would those projects that permit shifting of resources to and from private projects. To measure the "separation" variable, projects were categorized into four groups: (a) those that were not segregated at all from privately sponsored projects, (b) those which involved little separation of resources, (c) those with considerable separation, and (d) those for which the people and materials were kept completely apart from projects that were carried out with the firm's own funds.

Another variable that has been mentioned as being a possible determinant of the degree of substitutability/complementarity of public and private R & D funds for a given government sponsored project (and one for which we have obtained information) is the primary originator of the idea for the project. One might expect that Federally supported projects that were devised with some participation of the firm would tend to be more consonant with both the economic and technical objectives of privately supported projects and hence would be more likely to lead to "spinoff" of other projects than would those that were devised without any involvement of the firm's personnel.

Another variable that was considered to possibly affect the extent of substitutability/complementarity between public and private R & D for each project was whether or not the project involved research, as opposed to development. Research projects might be expected to lead to greater "spinoff" of private R & D ventures given that they are less specific in nature than are development efforts (though this may depend upon the extent that the firm's global R & D program is geared primarily at shorter range and less risky activities—which seems to be the trend for most firms).[41]

Another variable that was considered as a possible factor influencing the degree of substitutability/complementarity between government and private R & D at the project level is the firm's research intensity, which we measured as the ratio of total company funded R & D expenditures to sales for the firm. We might hypothesize that more research intensive firms may provide a more serendipitous atmosphere for government R & D to spinoff private R & D. An offsetting force, though, might be budget limitations which might prevent newly generated ideas from being exploited.

The last variable that was considered as a potential determinant of substitutability/complementarity between public and private R & D is the size of the firm's total private R & D budget. As was mentioned at the end of the previous section, it seems difficult to predict the effects of R & D size (which is positively related to firm size, though not proportionally) as such on the degree of substitutability/complementarity between government and private R & D. One might assert that after taking account factors which measure the degree of technological association of government and private R & D projects (e.g., measured by contribution of projects to the firms' technological objectives as above) as well as their economic association (e.g., measured by the extent of separation of resources employed on government supported projects and privately funded projects), absolute size should not matter. An alternative view might extend the neo-Schumpeterian argument to assert that large size might, for economies-of-scale reasons, be more conducive to complementarity, independent of project-specific factors. This is an additional proposition that we shall test.

The Evidence

Consistent with some of the more recent findings (including our own) regarding the issue of substitutability vs. complementarity of government and private R & D (which are based on different methodologies and do not differentiate R & D by function), in aggregate our data appear to indicate that the forces of complementarity between government and private expenditures overwhelm for energy-related R & D.

First, with regard to the net incremental impact of the projects on the firms' overall R & D programs, on average about 64.2 percent of the funds allocated to the projects would have been spent on R & D if the firms did not accept the projects.[42] Alternatively we can state that about 36 percent of the funds spent on the project represented an incremental gain to the firm's total R & D program, on average. We also note that for 15 out of the 40 projects for which data were provided, spinoff of R & D projects into which the firms invested private funds occurred. Finally, we note that in the absence of government support for the projects in over 50 percent of all cases (22 out of 41, to be precise), the firms would not have performed any of the work involved in the projects at all.[43] On the average, firms would have carried on about 17 percent of the work embodied in these projects if they had not received government support.[44]

Let us look now in some detail at some of the factors that seem to affect a major area of concern for public policy: whether a given government supported project is likely to spinoff projects in which the firm invests its own funds.

Suppose that the value to the firm of supporting additional R & D projects using private funds based on results of a given government sponsored project can be written as: $Y_1 = X_1\beta_1 + U_1$, and that the value of the event in which no spinoff occurs can be written as: $Y_2 = X_2\beta_2 + U_2$, where the X_i's are K element vectors of factors affecting the value or utility of each possibility to the firm, and the U_i's are random error terms. Let us denote the event that spinoff occurs as $Y = 1$. Similarly, denote the event that no spinoff occurs as $Y = 0$. Assuming that the firm makes the most valuable decision, we have

$$Y = 1 \Rightarrow Y_1 > Y_2 \text{ and}$$
$$Y = 0 \Rightarrow Y_2 > Y_1,$$

or alternatively

$$Y = 1 \Rightarrow U_2 - U_1 < X_1\beta_1 - X_2\beta_2.$$

Let us denote

$$\hat{u} = U_2 - U_1, \text{ and } (X_1\beta_1 - X_2\beta_2) = X\hat{\beta}.$$

Thus,

$Y = 1$ $\hat{u} < X\hat{\beta}$

$Y = 0$ otherwise.

Suppose that U follows the Sech^2 distribution:[45]

$f(\hat{u}) = e^{\hat{u}}/(1 + e^{\hat{u}})^2$ for $-\infty < \hat{u} < \infty$.

Then given X_n the ex ante probability that $Y = 1$ can be written as the cumulative:

$P_n = F(X_n\hat{\beta}) = \int_{-\infty}^{X_n\hat{\beta}} e^{\hat{u}}/(1 + e^{\hat{u}})^2 \, d\hat{u} = 1/1 + e^{-X_n\hat{\beta}}$.

Thus, the likelihood of the ith observation is:

$L_i = P_n$ if $Y_i = 1$

$L_i = (1 - P_n)$ if $Y_i = 0$.

This is the logit model for the problem.[46]

If we have a sample of N independent observations for which the event $Y = 1$ is observed N_1 times and which $Y = 0$ $N_2 = N - N_1$ times, the likelihood of the sample can be written as:

$$L1 = \Pi_{n=1}^{N_1} P_n \Pi_{n=n_1+1}^{N} (1 - P_n)$$
$$= \Pi_{n=1}^{N} P_n Y_n (1 - P_n)^{1-Y_n}$$
$$= \Pi_{n=1} (1/1 + e^{-X_n\beta})^{Y_n} (1/1 + e^{+X_n\beta})^{1-Y_n},$$

using the fact that Y_n is either zero or one.

In order to estimate the K parameter vector β we used the maximum likelihood procedure. Assuming various regularity conditions, maximum likelihood estimates (MLEs) are: (i) consistent, (ii) asymptotically efficient, (iii) asymptotically normal.[47] MLEs have also been found to have desirable properties even when applied to small samples in many circumstances for the logit model.[48]

The technique in essence involved choosing estimates to maximize the log of the likelihood function (which is equivalent to maximizing L_1). Taking the natural log of L_1 and then differentiating with respect to the β_k's, $k = 1, \ldots, K$, then we arrive at a set of K nonlinear equations for the K exogenous variables in the model.[49] To solve the system, we employ the Davidon-Fletcher-Powell method.[50]

In our discussion of government-industry interaction in the performance of energy-related R & D, we listed a number of factors that might serve as candidates for the X variables in the specification L_1. Of course these factors can by no means be expected to be the sole determinants of the nature of the interaction between government and private energy-related R & D, which may depend in many instances on institutional and/or political factors that are difficult or even impossible to quantify.[51] Nevertheless, they seem reasonable and most are in accord with the preconceptions of various authorities who have worked in this area. At any rate, in specifying the models, we noticed patterns of strong interdependence among some of these factors. For example, significant relationships were found between our measures of technological and economic integration. More specifically, projects were both ex ante and ex post less likely to contribute to the firm's technological objectives compared to expenditures of similar amounts on projects in which the firm invested its own funds were usually kept separate from projects that were privately financed.[52]

The "contribution to technological objectives" variables were also significantly related to a variable representing the identity of the party or parties most responsible for the generation of the idea for the project. In particular, projects whose ideas originated without any participation of the firm tended not to contribute to the firms' technological objectives.[53]

The strong interdependence among these three factors adversely affected the unconstrained estimation of L_1. In all specifications in which (either the ex ante or ex post) "contribution to technological objectives" variable was included with either the "idea" variable and/or the "separation" variable, the models lacked significance (i.e., their predictions were not better than a naive random choice model). When the "contribution to technological objectives" variable was excluded, however, the fits improved markedly. In general, we found that the idea variable (which may serve also as a proxy for the degree of technical and/or economic integration of public and private projects, in light of the above) tended to be the most significant determinants of substitutability/complementarity of government and private energy R & D. As was predicted in the previous section, firm R & D size as such apparently did not aid in explaining the substitutability/complementarity relationships. The distinction between research and development also did not appear to systematically influence the phenomena of interest.

We report in Table 2.8 the maximum likelihood estimates and associated t-values for three alternative models corresponding to L_1.[54] As can be inferred from the small changes in the values in the log likelihood functions, there is no significant explanatory power to be lost in proceeding from Model 1C to either 1B or 1A.

It is clear that as we indicated, the originator of the idea for the project (which perhaps also serves as a proxy for the integration of the work embodied in the project and privately funded projects) significantly affects the likelihood that any given government sponsored project will spinoff additional projects into which

Table 2.8.   MLE Estimates for *L*1 Based on 40 Projects:
Dependent Variable (*Y*ₙ) Equals 1 if Government Sponsored Project
Spun-Off R & D Projects in which the Firm Invested Its Own Funds;
Absolute *t* Values in Parentheses

Independent Variable	Model 1A	Model 1B	Model 1C
Constant term	−0.143	0.020	0.075
	(0.378)	(0.049)	(0.107)
Idea[a]	−1.466[c]	−1.483[c]	−1.504[c]
(binary)	(1.700)	(1.700)	(1.671)
Firm R & D to Sales Ratio	——	——	−0.036
			(0.196)
Separation[b]	——	−1.286	−1.283
(binary)		(1.092)	(1.090)
χ^2	5.966[d]	7.394[d]	7.404
Log of the Likelihood Function	−24.743	−24.029	−24.024

[a] This variable equals 1 if the firm was not involved in the generation of the idea for the project and equals zero otherwise.
[b] This variable equals 1 if the people and materials used on the project were kept completely separate from projects which were privately financed, and equals zero otherwise.
[c] Indicates coefficient is significant in a one-tailed test with $\alpha = 0.05$.
[d] Indicates coefficient is significant in a one-tailed test with $\alpha = 0.10$.

the firm invests private funds. We note that Model 1A predicts that probability that complementarity between government and private R & D in the form of "spinoffs" of privately funded R & D projects as a result of government R & D will be observed for a project whose idea did not originate (at least partially) within the firm is only about 0.17.

There is somewhat weaker evidence for the hypothesis that separation of resources adversely affects the likelihood that a given government sponsored project will lead to spinoff of private R & D, though the evidence is somewhat weaker than for the "idea" variable. The firm research intensity variable, however, does not appear to be a significant explanatory factor here, indicating perhaps that the serendipitous element that was mentioned previously is not adequately captured by this variable.

In total, an obvious policy implication of these results is that to the extent that spinoff of private R & D is an important objective of government funded R & D (in addition to other objectives), government officials should try to encourage the participation of private firms in the development of ideas for publicly sponsored projects and, at the same time, try to relax any severe restrictions regarding the amount of interaction permissible between the resources employed on existing government and privately supported projects.

Summary and Conclusions

Government support of energy-related R & D in private industry has risen at an extremely rapid pace over the past decade. In fact, the rate of growth of government sponsored energy-related R & D exceeds the rate of growth of all other budget functions; as shall be shown in chapter 3, this rate exceeds the rate of growth of inflation in R & D by a substantial margin.

The rationale for increased government involvement in energy R & D (as well as in R & D in general) that is usually adduced is that social returns exceed private returns; the private sector acting on its own will underinvest in energy R & D owing to risk aversion and/or limitations on appropriability of returns. Another motive that is often advanced is that expanding the nation's energy sources is essential to national security. Thus energy R & D is increasingly looked upon as analogous to defense R & D from a public policy point of view.

Despite the importance of the topic, little attention has been paid to analyzing the effects of government involvement in energy R & D, in particular on measuring the extent of substitutability/complementarity of public and private R & D expenditures. Economists have examined the issue of substitutability/complementarity for R & D as a whole, however. Recent studies have tended to suggest that on the whole, complementarity obtains for public and private R & D expenditures, though the methodologies used seem to be open to question on a number of grounds, as was discussed in some detail.

The approach of this chapter differs from previous studies (including the work of the previous chapter) that have examined the relationship between public and private R & D expenditures in at least two respects. First, we have been able to address the problem from a more meaningful level of disaggregation. Second, because we have both ex ante and ex post data, we are able to address richer counterfactual statements. For example, the reduced-form investment approaches using historical data series only provide us with information on trends in the ex post responses of firms to changes in government policy variables; our willingness to accept the predictions gleaned from these models depends upon our confidence in their ability to correctly specify the relevant variables and the lag structures pertaining to these variables. Our data have permitted us to capture both the intentional responses of firms as well as actual responses.

On the whole, the evidence we uncovered for energy R & D is consistent with the evidence of recent studies for aggregate R & D that supports the notion that government funded energy R & D complements privately funded energy R & D. On the average, if a firm did not accept a given government R & D project, only about 64 percent of the funds spent on a project would have been spent by the firm on R & D. Also, 15 out of 40 government supported projects for which data were provided spun-off further R & D projects into which the firms invested private funds. Finally, on the average, firms would have carried out only about

17 percent of the work embodied in these projects in the absence of government support.

In addition to examining the magnitude of the substitutability/complementarity relationship, we also looked at some of the determinants of substitutability/complementarity. A key policy implication of the analysis is that to the extent that spinoff of private R & D efforts is an important objective of the government when allocating energy R & D contracts to the private sector, the government should do its utmost to encourage the participation of private firms in the generation of ideas for R & D projects, and should not enforce any rigid requirements on the separation of the resources applied to the projects from privately supported projects, once contracts are awarded.

3

Inflation and R & D Expenditures

Introduction

Since antiquity, the price mechanism has been lauded as a superior means of allocating the resources of society. F. Hayek, an outspoken advocate of the system, states:

> In a system where the knowledge of the relevant facts is dispersed among many people, prices can act to coordinate the separate actions of different people in the same way as subjective values help in the individual to coordinate the parts of his plan. . . . the most significant fact about this system is the economy of knowledge with which it operates, or how little the individual participants need to know in order to be able to take the right action.[1]

For the price mechanism to work, though, the individual agents in the economy must be fully cognizant of the true prices and costs that prevail. In the past, policy makers have opted for a number of alternative proxies for the price of R & D as a guide to the allocation of resources aimed at enhancing technological change. Unfortunately, none of the indices used for industrial R & D is grounded in economic theory. To the extent that the decisions of policy makers regarding R & D expenditures in both the private and public sectors have been and will continue to be guided by perceived changes in the costs of performing R & D relative to other activities,[2] it seems important that adequate price deflators for R & D be made available.

In pursuit of this objective, we will present a number of alternative indices that we have developed for a number of U.S. manufacturing industries for the recent period 1969–79, using project level and company level data that we have collected from some of the leading R & D performers in the U.S. In the next section, we will discuss some of the previous efforts to construct price deflators for R & D that have appeared. We will then provide some discussion of the methodology we use in relation to the economic theory of price index literature. An analysis of the project specific data follows, which will be proceeded by a discussion of the company-specific data. Next, the various indices will be compared and contrasted with alternative measures of the rate of inflation in R & D that have been used. The chapter concludes with a discussion of the implications of our findings for public policy.

Review of Previous Work

To date, very little work has been done in developing price deflators for R & D. In fact, there is at present no official R & D deflator available. The GNP deflator is used by government agencies for this purpose. Two of the main attempts to construct R & D price indices for the U.S. in the past decade are those of H. Milton (1966, 1972) and S. Jaffe (1972).

Milton's "cost of research" index was constructed by taking annual R & D costs of a sample of organizations (17 in the 1966 study and 15 in the 1972 update) as a percentage of the organizations' total scientific and technical employment.[3] This index was expected to "provide a measure of the cost of the technical man-years supported by the R & D funds expended in the U.S. relative to the R & D man-year effort achieved."[4] Unfortunately, the index provides no means of holding the variable "man-year effort achieved" constant, and hence cannot differentiate between the effects of inflation per se and changing levels of real effort (involving both human and nonhuman capital) devoted to R & D. Milton recognizes this shortcoming. As she states, the cost of research index cannot measure the effectiveness of a given man-year of effort "either in terms of amount or quality . . . the numbers of trained support personnel and augmentation by machines affect the productivity of the average technical man-year to an unmeasured degree."[5] Clearly, this index cannot be considered as an adequate price deflator, for as Griliches notes, "it does not hold either the composition of the research and development labor force or the quantity of other inputs purchased constant."[6] Nevertheless, despite these problems, it has been used as the basis for the annual price index for R & D in the U.S. published by the Battelle Memorial Institute.

Jaffe's approach, in contrast to Milton's, has some theoretical appeal, although the index he computes is somewhat limited in scope. Using NSF data on expenditures for R & D at universities and colleges (excluding Federally funded research and development centers) he developed a Laspeyres price index for academic R & D for the period 1961–71 using base period weights from 1967. His main conclusion of interest here is that in the light of his computed index, the rate of growth of real R & D expenditures from 1961–71 is overestimated, when nominal expenditures are deflated by aggregate measures such as the GNP deflator.[7] Three limitations of using his approach for determining the rate of inflation in industrial R & D in the U.S. might be mentioned.

Perhaps the least serious problem is his use of the Laspeyres method, which among other potential problems (to be discussed in detail in the following section) ignores substitution effects. Second, there is the matter of incomplete coverage of various costs of R & D. For as he notes, only direct costs, including personnel compensation, purchases of small and expendable equipment, and purchases of supplies charged to projects are included.[8] The services of separately budgeted capital equipment and expensive equipment are ignored, as are overhead charges. Finally, the index only relates to academic R & D, most of which is classified as basic research.[9]

This brief review covers the brunt of previous efforts to construct R & D price deflators over the past decade.[10] Prior to examining our data, let us review the theory underlying the price indices we will be presenting.

Some Relevant Theory

At the outset it should be mentioned that the deflators which we develop refer to the costs of R & D inputs. Although a few attempts have been made to quantify the outputs of R & D for firms,[11] aggregative data in the form required to construct output deflators for a variety of industries have not been forthcoming. Following Fisher and Shell's (1972) remarks,[12] a properly construed R & D input price deflator should be able to answer the following question: Given two situations A and B, what R & D budget size would be required to yield a predetermined level of "R & D output" given B's prices, as opposed to A's prices and budget level? If a specific functional form for the "R & D production function" is assumed, one can in principle answer this question precisely by solving the following mathematical programming problem:

$$\text{minimize } \Sigma_{i=1}^{n} W_i X_i \text{ subject to: } f(X_1, \ldots ,X_n) = y,$$

where
$$W_i \equiv \text{the price of input } i$$
$$X_i \equiv \text{the quantity of input } i$$
$$y \equiv \text{the designated output level}$$
$$f \equiv \text{the R \& D production function.}$$

Let us write the minimum value achieved by the objective function with price vector W^A and price vector W^B as $C^*(W^A,y)$ and $C^*(W^B,y)$ respectively. The price index for comparing situation A with situation B can be written as:

$$P^* = C^*(W^B,\bar{y})/C^*(W^A,\bar{y}).$$

For example, with a Cobb-Douglas production function of the form

$$y = A \ \Pi_{i=1}^{n} X_i^{\alpha_i},$$

the price index can be solved using the above procedure as:

$$P^* = \Pi_{i=1}^n (W_i^B/W_i^A)^{\alpha_i}.$$

Analogously, for the CES production function of the form:

$$y = \{A \, \Sigma_{i=1}^n \delta_i X_i^{-\rho}\}^{-1/\rho},$$

the index would be

$$P^* = \Sigma_{i=1}^n(\delta_i \, W_i^{B_{1-\sigma}})^{1/1-\sigma}/\Sigma_{i=1}^n (\delta_i W_i^{A_{1-\sigma}})^{1/1-\sigma},$$

where $\sigma = 1/1 + p$ is the elasticity of substitution. For a translog cost function of the form

$$\ln C = \alpha_0 + \Sigma_{i=1}^n \alpha_i \ln W_i + 1/2 \, \Sigma_{i=1}^n \Sigma_{j=1}^n \alpha_{ij} \ln W_i \ln W_j,$$

with $\Sigma_i \alpha_i + 1, \; \alpha_{ij} = \alpha_{ji}, \; \Sigma_j \alpha_{ij} = 0,$

(which provides a second order differential approximation to an arbitrary twice continuously differentiable unit cost function), the price index can be written as[13]

$$P^* = \Pi_{i=1}^n (W_i^B/W_i^A)^{1/2} (W_i^A X_i^A/\Sigma_i W_i^A X_i^A) + (W_i^B X_i^B/\Sigma_i W_i^B X_i^B).$$

Unfortunately, using all of the functional forms above to estimate an exact cost of research indices was precluded owing to a lack of data necessary to estimate the various parameters with confidence. Instead, we have had to limit our indices to the Cobb-Douglas, Laspeyres (L) and Paasche (P) forms, where

$$L = \Sigma_i W_i^1 X_i^0/\Sigma_i W_i^0 X_i^0 \, ; \; P = \Sigma_i W_i^1 X_i^1/\Sigma_i W_i^0 X_i^1 \, ,$$

where $W_i^1 \equiv$ price of input i in the current period
$W_i^0 \equiv$ price of input i in the base period
$X_i^1 \equiv$ quantity of input i in the current period
$X_i^0 \equiv$ quantity of input i in the base period.

The (L) and (P) forms can be shown to represent exact cost of R & D indices if the technology is of the fixed coefficients form.[14] If we are unwilling to maintain this assumption due to preconceptions about definite substitution possibilities amongst inputs, we can at least show that (L) and (P) forms provide certain bounds on the true cost of research index, assuming a convex technology set for R & D outputs. In particular, we can demonstrate that the Laspeyres index provides an upper bound to the true cost of research index if the reference isoquant is that of the base period (i.e., period 0). First, assuming cost minimization behavior, we write the true cost of R & D index with the output level y^0 as

$$P^* = C^*(W^1, \bar{y}^0) / C^*(W^0, \bar{y}^0) = C^*(W^1, \bar{y}^0) / \Sigma_i W_i^0 X_i^0.$$

The Laspeyres index again is

$$\Sigma_i W_i^1 X_i^0 / \Sigma_i W_i^0 X_i^0.$$

Assuming convex isoquants, then, it is clear that

$$\Sigma_{i=1}^n W_i^1 X_i^0 \geq C^*(W^1, y^0),$$

so the assertion holds. Diagrammatically, the argument is shown below for the case of two inputs. Originally, the optimizing firm facing relative prices given by P_0 and a budget constraint of C operates at point A using the combination (X_1^0, X_2^0) to produce y^0. If we were to change relative prices to, for example, the ratio given by P_1 and at the same time subsidize the firm to maintain the factor inputs (X_1^0, X_2^0) as a means of preserving output at y^0 we would be overcompensating for the price change, since the new budget at A, $W_1^1 X_1^0 + W_2^1 X_2^0$ is clearly greater than that required, with the new optimal combination of inputs at B.

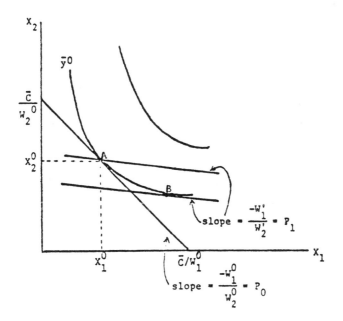

A symmetric argument can be made to demonstrate that the Paasche index provides a lower bound to the true cost of R & D index when the reference output level is that of the current period. It might be mentioned that in the case of a homothetic technology for the R & D production function, we can state that the true price index is independent of the reference level of output.[15] In this case, the Laspeyres and Paasche indices provide upper and lower bounds for the cost of R & D index regardless of the output level used as the reference point. The assumption of homotheticity may be overly restrictive, however.

In the presence of rapid changes in the technology of R & D labs over time which alters the relative productivity of various inputs (e.g., due to the gains that have been made in instrumentation) of all the indices which we consider, one might argue that the Paasche form is the most relevant as a guide to planning. For in this case, the Paasche index, which focuses on the current technology, maintains its "lower boundedness" property, whereas the Laspeyres index will lack a known relationship to an exact deflator for R & D based on today's technology.[16] Use of the Cobb-Douglas index with fixed weights is tantamount to assuming an unchanging technology for R & D output over time.

Owing to the nature of our data, however, the main indices we employ have the Laspeyres form. According to the Organization for Economic Cooperation and Development, the Laspeyres form has been preferred in experimental work pertaining to R & D price deflators in other countries.[17]

This concludes our review. In light of the discussion above, the reader should bear in mind the various assumptions underlying the indices which we develop in the following sections.

Project-Level Indices

A most direct approach to computing R & D price indices which we followed involved the selection of a set of actual R & D projects that were performed in a firm, and the generation of deflators based on actual costs incurred on the various disaggregated inputs to the projects. This approach seems reasonable, and provides a check or standard for evaluating more aggregative measures of inflation in R & D.

To demonstrate the potential utility of this approach, in this section we will present Laspeyres and Paasche indices that we have developed for two large scale R & D projects that were performed in an actual firm in private industry.

A General Description of the Data

The data used to construct the price indices were obtained from one of the leading chemical producers in the U.S., with 1980 sales of the corporation exceeding $1 billion. The firm is furthermore one of the more research intensive of all firms in its two-digit SIC industry.

In amassing the necessary data, a senior R & D executive of the corporation was consulted, who for the purpose of this study, chose two representative R & D projects[18] in the firm's most research intensive line of business, agricultural chemicals[19]; a project which was performed over the period 1965 to 1977 was selected as the basis of the Laspeyres index calculations (project L), while a project performed between 1969 and 1982 served as the basis of the Paasche index (project P). For purposes of comparability, both indices were constructed to reflect price changes over the period 1969 to 1979. Project leaders, analytical scientists, and accounting personnel were met with on various occasions to collect the necessary data. In order to maintain confidentiality, we agreed to refrain from identifying the firm and the projects by name.

With regard to the nature of the R & D projects, both concern the development of herbicides. Although the projects were conducted at different times, the R & D activities captured by the indices are of a similar nature. The main R & D activities that are represented by each of the projects include process and plant design, toxicology and residue analyses, and field development studies.

The actual histories of the projects follow somewhat different paths (to be discussed next), although in both cases, the anatomy of the innovation process followed a common schemata (which we were told is the general course followed by R & D in herbicides, at least at this corporation). The first stage of herbicidal innovation in this schemata involves the *conception and synthesis of new compounds* by synthesis chemists who may be from outside or inside the organization. The next stage involves the *screening of the new compounds for their overall herbicidal utility* by a group of the firm's chemists and applied biologists. At this stage, only the general characteristics of the new compounds are subject to intense scrutiny. Third, once herbicidal utility is determined, the compounds are transferred to another group of chemists and biologists to determine *herbicidal selectivity,* measured by examining the specific crops/weeds that are immune to their treatments. The final stage is that of *developement* which embraces the traditional (NSF) definition of development as well as the stages subsequent to development in the anatomy of product innovation schemata of Mansfield et al. (1971, 1977)[20] and includes such activities as process and plant design, toxicology studies, and crop and soil residue analyses, as well as field and market studies. One of the increasingly costly elements of this stage involves the adaptation of the product to satisfy the stringent requirements of the Environmental Protection Agency[21] as well as the registration requirements of other countries. In both of the projects for which we derived price indices, it is primarily the final stage of development that is reflected in the analyses. Let us now turn to the projects themselves.

The Laspeyres Project

To calculate the Laspeyres index we estimated the cost of project L's inputs in 1969 based on 1979 prices relative to the costs actually incurred in 1969, which were $588,000.

The services of approximately 15 full-time scientist-engineers and 11 supporting personnel were allocated to the project in this year (1969).

In constructing the Laspeyres index, an attempt was made to disaggregate the costs of the various inputs as thoroughly as possible, while at the same time providing completeness of coverage. The actual costs that could be identified fall into five categories: (a) services of scientist-engineers, (b) services of supporting personnel, (c) elementary chemical or raw material inputs, (d) services of durable capital, and (e) other direct and outside services.[22] The scientist-engineer category was further broken down to account for differences in level of education attainment, differences in academic discipline, and differences in experience amongst individuals.[23] The Laspeyres indices computed for the various inputs as well as for the project as a whole are shown in Table 3.1.

As can be clearly seen, the rate of increase in the cost of performing R & D exceeds the level that would be implied by the GNP deflator (which was 188), which provides the standard for the official calculations of real R & D expenditures. The only item that seems to track the GNP deflator at all closely is the compensation of scientist-engineers, before accounting for fringe benefits.

The Paasche Project

To estimate the Paasche index for the firm, we took the ratio of the total R & D costs incurred in 1979 (which were $2,649,000) for project P, relative to the computed cost of the same inputs used in 1979, but at 1969 prices.

Approximately 26 full-time scientist-engineers and 17 supporting personnel were engaged in the R & D work in 1979. In computing the Paasche index, costs were disaggregated into the following six categories: (a) services of scientist-engineers, (b) services of supporting personnel, (c) elementary chemical or raw material inputs, (d) services of durable capital, (e) services of key analytical equipment, and (f) other direct and outside services. As with the Laspeyres project, the scientist-engineer category was disaggregated to account for different levels of educational attainment, differences in academic discipline, and differences in work experience among the personnel. The analyses relevant to (e) reflect the costs imputed to the project of performing nuclear magnetic resonance spectrometry which was crucial to the outcome of the R & D. Items (a) through (e) account for 70 percent of the total R & D costs allocated to the project in 1979.[24] The Paasche index for the project (calculated for the period 1969 to 1979) as well as the relative price changes of the various inputs are shown in Table 3.2.

Table 3.1. Laspeyres Price Index for the 1969 Project and Price Indexes for
Individual Inputs (1969 = 100)

Total	Scientist-[a] Engineers	Scientist-[b] Engineers	Supporting Personnel	Raw Material Inputs	Capital Services	Other
217	192	216	209	186	318	192

[a]Excludes fringe benefits   [b]Includes fringe benefits

Table 3.2. Paasche Price Index for the 1979 Project and Price Changes for
Individual Inputs (1969 = 100)

Total	Scientist-[a] Engineers	Scientist-[b] Engineers	Supporting Personnel	Raw Material Inputs	Capital Services	Analyses	Other
206	194	218	209	154	318	313	186

[a] Excludes fringe benefits   [b] Includes fringe benefits

It is evident that the overall rate of change in cost is fairly similar here to
the rate shown for the Laspeyres project, and clearly exceeds the rate implied by
the GNP deflator (which was 188). As with the Laspeyres project, the components
that appear to be increasing in cost most rapidly relative to the GNP deflator include the services of durable capital equipment, supporting personnel, and scientist-
engineers (only to the extent that fringe benefits are included in the estimation).
The costs of analyses also appear to be rising at a precipitous rate.[25] Our project-
level data thus seem to provide a certain amount of first-hand evidence against
the validity of the official practice of using the GNP deflator in determining real
allocations to R & D, at least for this firm.

Company-Level Indices

An alternative procedure to the "representative project" approach that was considered for developing R & D price deflators for industry involved looking at the
total R & D programs of a sample of firms over a period of time. This approach
abstracts from variations among individual projects and hence conceals a great
deal of information that is potentially of interest. Still, it may be useful for managers
and policy makers in their decisions regarding aggregate allocations of innovative
activities. In this section, we highlight some of the important findings of a study
in which Laspeyres indices were developed for R & D inputs as well as for inputs

to other stages of the innovation process for the firms over the period 1969–79. In the study we used company-level data to compute industry indices of the Laspeyres and Cobb-Douglas forms for R & D inputs and Laspeyres indices for the inputs to the stages of the innovation process. The discussion here will be divided into three parts. First, a short description of the sample will be provided. Next, the industry level price indices for the inputs to R & D will be presented. Finally, the deflators for the stages of innovation for the various industries studies will be shown.

Description of the Data

The data used in the analysis were obtained from firms operating in eight industries (which account for approximately one-half of all company financed R & D in the U.S.): chemicals, petroleum, electrical equipment, primary metals, fabricated metal products, rubber, textiles, and stone, clay and glass.

In amassing our primary source material, we originally contacted the senior R & D executives of a large number of firms in these industries. Wherever possible, we arranged for a series of meetings in which the problems could be discussed thoroughly, and we attempted to obtain their assistance in the collection of data. Assurances were given that the data would be treated with the strictest confidentiality and that the firms participating would not be identified. In a few cases, the firms' accounting records did not contain sufficient historical detail to complete the study;[26] in a few other cases, managers were unwilling to participate because of their feelings in regard to the costliness of performing the detailed calculations in the presence of serious budgetary cutbacks. Nevertheless, a total of thirty-two firms ultimately provided us with useful information for the study. Though the final sample contains both large and small firms, it contains a substantial proportion of the R & D carried out in the various industries,[27] as indicated by the fact that the firms included account for approximately one-ninth of all company financed R & D in the U.S.

Price Deflators for R & D, 1969–79

In deriving the R & D price indices, we obtained or computed from each firm a Laspeyres index which shows how much more it would have cost the firm in 1979 to hire the same R & D inputs as were hired in 1969. For each industry, a weighted mean of the component firms' indices was then calculated, using the R & D expenditures of the firms as relevant weights.[28] The results follow in Table 3.3. We also obtained in many cases data from each firm concerning the price increases over the period 1969–79 for five types of R & D inputs: wages of scientist-engineers, wages of supporting personnel, materials and supplies, services of R & D plant and equipment, and other inputs. Using these data, we also constructed exact price indices for R & D assuming the production function is Cobb-Douglas, recognizing the potential biases of the Laspeyres form due to its neglect of potential substitution among inputs. To compute the Cobb-Douglas indices the various input price relatives were weighted by two alternative measures of α_i (the share of total costs devoted to input i); one measure was derived from a study by L. Goldberg (1978)[29]. The other is based on NSF (1972) data.[30] The results are also shown in Table 3.3. It should be mentioned that not all of the firms in the sample provided data concerning price increases for each of the individual inputs, and thus the Cobb-Douglas results are not fully comparable with the Laspeyres indices computed for R & D.

Looking first at the Laspeyres indices, it is apparent that the rate of inflation for R & D is not uniform across industries. In fact, the rate of inflation appears to be highest in fabricated metal products, chemicals, and petroleum, and lowest in electrical equipment. A striking feature of these aggregative results is that the Laspeyres price index computed for the chemical industry (222) is very similar to the Laspeyres index for the chemical project (217), which was developed in the previous section and is almost identical to the index calculated for the combined items (a) scientist-engineer compensation (after accounting for fringe benefits), (b) supporting personnel compensation, (c) raw material inputs, and

Table 3.3. 1979 Laspeyres Price Index for R & D Inputs and
R & D Price Indices Based on the Cobb-Douglas Production Function
(1969 = 100)

Industry	Laspeyres Price Index	Cobb-Douglas Index — Estimates of α_i Based on Data from	
		Goldberg (1978)	National Science Foundation (1972)
Chemicals	222	217	214
Petroleum	222	218	218
Electrical Equipment	183	190	191
Primary metals	205	205[a]	204
Fabricated metal products	248	222	218
Rubber	209	206	205
Stone, clay and glass	205	183	183
Textiles	200	220[b]	222
Mean[c]	198	200	200

[a] The mean of Goldberg's figures for ferrous metals and nonferrous metals is used.

[b] The percentages of cost for scientists and engineers, support personnel, and materials and supplies come from National Science Foundation (1972). The balance is split between "services of R & D plant and equipment" and "other" in proportion to the mean percent in each of these categories as indicated by Goldberg.

[c] Each industry's price index is weighted by its 1969 R & D expenditures.

(d) capital services for the Laspeyres project (which account for 80 percent of the expenditures allocated to the project) and which together yield a Laspeyres index of 223.

With regard to the rate of inflation of the individual inputs to R & D, the wages of support personnel seem to have risen most rapidly. The OECD has found that this holds in other countries as well.[31] Materials and supplies have also experienced rapid cost increases overall, perhaps reflecting increased expenses on hydrocarbon based inputs. There are, however, important variations across industries. The costs of materials and supplies have risen more slowly than those of other inputs in the chemical industry. This is entirely consistent with our project level findings. Also consistent with the project level results, we note that the wages of scientists and engineers have risen at a relatively slow rate (compared to some of the other inputs). This may be due to the fairly slow growth in the demand for scientists and engineers and to their relatively bountiful supplies.[32]

Turning now to the Cobb-Douglas indices, it is readily apparent that regardless of which set of α_i is used, the final results are very similar to the Laspeyres results.[33] The interindustry differences are somewhat smaller for the Cobb-Douglas indices, though. Once again, the R & D price index tends to be relatively high in fabricated metal products, chemicals, and petroleum, and relatively low in electrical equipment. Looking in more detail at the chemical industry, it is apparent that the two Cobb-Douglas indices are quite similar to the Paasche index computed for the chemical project in the previous section,[34] and are somewhat smaller than the project and company level Laspeyres indices for chemicals. It is clear that the R & D price index for chemicals, regardless of the method chosen for computation, significantly exceeds the GNP deflator, as is the case for most of the industries here.[35] Let us now turn to the indices for the various stages of the product innovation process.

Laspeyres Price Indices for the Stages of the Product Innovation Process, 1969–79

The product innovation process can be viewed as encompassing the activities undertaken from the onset of exploratory work with a particular new product in mind to the time at which the new product is available for sale and delivery on the market.[36] Again, it can be subdivided into the six-stage format (as defined in n. 20) of applied research, preparation of project requirements and basic specifications, pilot plant or prototype design and construction, tooling and construction of manufacturing facilities, manufacturing startup, and marketing startup. Recall that only the first three of these stages would be treated as R & D by the usual definitions. However, product innovation involves much more than R & D. In fact, in a detailed study of chemical innovations, it was found that on average, about 61 percent of all of the costs of innovation were not categorizable as R & D.[37]

As with the Laspeyres price deflators for R & D, we obtained or computed for each of the firms of the sample a Laspeyres index which measures the increased costs of hiring the inputs of 1969 in 1979 for each of the six stages of the innovation process, as well as an index for the innovation process as a whole. The weighted means of the indices were calculated for each industry, using the component firms' R & D expenditures as the relevant weights.[38] The results are shown in Table 3.4.

Apparently the rate of inflation for the entire innovative process is slightly higher than for R & D alone. As with R & D, the rate of inflation is highest in the fabricated metals, chemical, and petroleum industries and is lowest in the electrical equipment industry. Also, as with the R & D indices, the rate of increase in the price of innovation exceeds that implied by the GNP deflator for all industries, with the exception of electrical equipment. Looking now at the individual stages of innovation, it appears that the highest average rates of inflation are achieved by the prototype or pilot plant stage and the tooling and construction of manufacturing facilities stage,[39] although we note that some variability exists across industries.

Table 3.4. Laspeyres Price Index for Inputs in the Innovative Process (and Individual Stages of This Process), Eight Industries, 1979 (1969 = 100)

Industry	Entire Innovative process[a]	Applied Research	Specifications	Prototype or Pilot Plant	Tooling and Manufacturing Facilities	Manufacturing Startup	Marketing Startup
Chemicals	223	204	229	218	217	224	225
Petroleum	228	217	201	239	254	225	232
Electrical equipment	186	182	182	200	200	184	185
Primary metals	210	249	234	236	226	231	227
Fabricated metal products	275	202	201	300	252	300	300
Rubber	200	——	——	——	——	——	——
Stone, clay and glass	195	185	178	195	232	220	163
Textiles	220	200	175	225	250	200	175
Mean[b]	201	193	196	211	211	202	201

[a] The price index in this column is not comparable with the indices for individual stages of the innovative process in other columns because some firms provided data for individual stages but not enough data to calculate an index for all stages combined. Other firms provided data for all stages combined, but none for some individual stages. Also, other columns are not entirely comparable with one another for this reason.

[b] Each industry's price index is weighted by its 1969 R & D expenditures.

Table 3.5. GDP Deflator and Estimate of R & D Deflator —
Both for 1975, Ten Countries
$(1967 = 100)$[a]

Country	R & D Deflator	GDP Deflator
Germany	177	156
United Kingdom	227	218
Netherlands	224	199
Sweden	177	171
Switzerland	176	164
Belgium	222	168
Italy	293	235
Canada	216	187
Japan	249	195
France	232	196

[a] The figures for the Netherlands, Italy, Canada, Japan, and France pertain to 1976, not 1975.
Source: Organization for Economic Cooperation and Development (1979).

Summary and Conclusions

This chapter has been devoted to estimating price deflators for R & D in the U.S. On the basis of the data that we have collected from firms that as a group account for about one-ninth of all company financed R & D in the U.S., the appropriateness of the usual proxies is brought into question. For example, over the period 1969 to 1979, the "cost of research index" published by the Battelle Memorial Institute increased by 84 percent.[40] Over the same period, the GNP deflator increased by about 88 percent. Though the price indices for R & D inputs that we developed vary across industries, in almost all of the industries studied here, the price increase for R & D inputs exceeded both of these rates. Official U.S. statistics on deflated R & D expenditures appear to overestimate the increase in real R & D expenditures for the industries we studied, taken as a group, from 1969 to 1979. Real R & D expenditures increased by approximately 7 percent based on the GNP deflator and by a slightly larger margin based on the Battelle deflator, but by less than 1 percent on the basis of our indices for R & D inputs.

To conclude, to the extent that policy makers in the U.S. wish to make wise decisions in the future in regard to the allocation of resources to technological change, it seems important that studies of this sort be expanded and replicated for future years. Hopefully, in the future data can be uncovered by researchers to permit the development of indices based on less restrictive assumptions, such as the "superlative" translog form, and that allow for exigencies such as technological change in the R & D process itself. Our approach of using company and project-level data on R & D inputs to develop R & D and innovation price deflators might also be applied by other countries interested in examining the effects of inflation on the innovative efforts of industry. At present, very rough Laspeyres price indices for R & D have been computed for a number of OECD countries, which appear to indicate that the rate of inflation in R & D has exceeded the economy-wide rate of inflation, measured by the GNP deflator, which is also true in our analysis. This is shown in Table 3.5. Again, these efforts are quite crude, since proxy series for the relevant inputs, rather than actual R & D inputs of firms are used in the computations. Much more can and should be done.

4

Conclusion

Economic researchers using fairly diverse methodologies seem to have reached a consensus regarding the importance of R & D in promoting the rate of productivity growth in the U.S. In addition, downturns in the rate of growth of investment in R & D has been cited as one of the salient factors underlying poor performance of the economy. In evaluating the various options that have been suggested for stimulating private R & D in an effort to reverse the recent economic setbacks, policy makers ought to be aware of the key factors that appear to determine the costs of the private sector's allocation of resources to industrial R & D and innovation. A major portion of this work has been devoted to this issue. We believe that our studies shed new light on this topic.

Previous attempts at modeling the R & D investment decision process have largely ignored the firm's financial constraints entirely. This may be unwarranted.

Consistent with earlier, less general formulations, we found that the firm's internal financing capabilities was a significant determinant of interfirm differences in R & D expenditures. We also uncovered some possible evidence of aversion to external financing of R & D. An immediate policy implication of this finding is that the Reagan administration tax credit to R & D, to the extent that it favorably influences the cash flow of affected firms, may have its desired effects of stimulating private R & D. An obvious caveat, though, which we mentioned at the outset, is that as with the Canadian experience, the government may not be able adequately to prevent firms from taking advantage of the tax credit by arbitrarily redefining activities under the rubric of R & D that are not typically thought of as R & D per se. Another shortcoming of the tax credit[1] is that R & D investments with potentially high social payoffs will not necessarily be forthcoming since for "firms that can appropriate little of the social returns from new technologies, R & D would still be unprofitable, even if the tax credit existed."

Government expenditures on civilian R & D may also be warranted when the private sector displays a high degree of risk aversion toward the funding of projects with high potential payoffs. An argument against government financing of R & D activities conducted by the private sector is that public expenditures may merely substitute for private expenditures. This is the familiar "crowding out" principle. Consistent with a number of recent studies using similar aggregative data, the results of chapter 1 do not bode well for the crowding out hypothesis.

In chapter 2 we also presented more disaggregative evidence against the hypothesis of crowding out of private expenditures by public expenditures for Federally funded energy-related R & D projects conducted in private research labs at various times between 1973 and 1980. For our sample on average, if firms did not accept the projects, only about 64 percent of the funds expended on the projects would have been allocated to R & D. Also, we showed that in 15 out of the 40 government supported projects for which data were provided, spinoff of private R & D projects occurred. Lastly, we showed that on average, in the absence of government support, only 17 percent of the work involved in the projects would have been performed by the firms themselves. Clearly, these results do not appear to support the notion that government expenditures are a strong deterrent of private R & D.

In addition to demonstrating that government R & D expenditures may on the whole positively impact upon private R & D expenditures, we also uncovered some firsthand evidence regarding factors that appear to promote complementarity between public and private R & D. In particular, we showed that spinoff of private R & D by government R & D may be enhanced by the practices of (i) encouraging the private sector's participation in the generation of ideas for R & D projects, and (ii) avoiding the imposition of regulations which require that resources devoted to government sponsored projects be divorced from privately supported projects.

In formulating a rational and comprehensive program aimed at reversing the slow growth trend of private R & D expenditures to boost productivity growth, it seems important that policy makers be aware of the extent of the problem. Using both case studies as well as production function approaches, researchers have shown that in the past, there has been an apparent underinvestment in civilian technology in the U.S., reflected by fairly high returns to investment in R & D, relative to other activities.[2] Little has been done in recent years to determine the persistence of this phenomena. In particular, no satisfactory evidence has been forthcoming to determine whether or not nominal R & D expenditures in the U.S. have kept pace with inflation. In the past, a number of indexes have been used as deflators for the conversion of nominal R & D expenditures into real magnitudes. Unfortunately, all of the various indexes are based on rather imperfect proxy series. In chapter 3 we attempted to develop more comprehensive price deflators for R & D that are based on actual data (and not proxy series) obtained from a number of firms. Our analysis apparently indicates that inflation may have adversely affected real allocations to R & D over the decade 1969-79 to a greater extent than has hitherto been recognized.

Before closing, a number of potential areas for future work should be indicated. First, it might be fruitful to reestimate the aggregative investment model of chapter

1 for the firms in our sample for alternative years; the researcher wishing to update the study may find that the task has been rendered somewhat complex, however, as a result of a recent merger activity that has affected a number of the firms in the sample.

Our approach to examining the effects of government expenditures on energy R & D might also be usefully adopted in a study dealing with other areas of civilian R & D.

Finally, the alternative methodologies used for the construction of R & D price deflators ought to be extended and replicated for future years in the U.S. and perhaps in other countries as well.

Notes

Chapter 1

1. See, e.g., S. Myers, "Interactions of Corporate Financing and Investment Decisions—Implications for Capital Budgeting," *Journal of Finance* 29 (March, 1974), pp. 1–25; C. Haley and L. Schall, *The Theory of Financial Decisions* (New York: McGraw-Hill, 1979).

2. M. Hellwig, "Bankruptcy, Limited Liability and the Modigliani-Miller Theorem," *American Economic Review* 71 (March, 1981), pp. 155–70.

3. The endogenous variables as well as nonratio financial variables were measured in millions of dollars.

4. Dhrymes and Kurz (1967, 431–32).

5. McCabe (1979, 120).

6. For at equilibrium, we know that the marginal rate of return on alternative uses of funds (R & D expenditures, investment and dividends) must equal the marginal cost of financing. Linearizing arbitrary rate of return functions, we might obtain, as in Grabowski and Mueller, the following:

   (1) $\quad MRR_{RD_i} = \alpha_0 + \alpha_1 RD_i + \alpha_2 I_i + \alpha_3 X_{PD_i}$,

   (2) $\quad MRR_I \;\;= \beta_0 + \beta_1 RD_i + \beta_2 I_i + \beta_3 X_{I_i}$,

   (3) $\quad MRR_{DIV} = \gamma_0 + \gamma_1 RD_i + \gamma_2 I_i + \gamma_3 X_{DIV_i}$,

   with the variables defined as $RD_i \rightarrow$ company funded R & D expenditures by firm i; $I_i \rightarrow$ capital expenditures for firm i; $DIV_i \rightarrow$ dividend payments of firm i; $EF_i \rightarrow$ external financing of firm i; $X_{RD_i}, X_{I_i}, X_{DIV_i}, X_{EF} \rightarrow$ vectors of exogenous determinants of RD_i, I_i, DIV_i, and EF_i, where a priori one might expect α_1, β_2 and γ_3 to be less than zero due to diminishing returns. The equation representing the marginal cost of financing we use is of the form:

   (4) $\quad MCF \;\;= \delta_0 NE_i + \delta_1 ND_i$,

   where $\quad NE \equiv$ new equity issued by firm i
   $\qquad\quad ND \equiv$ new debt issued by firm i.

   This equation could be obtained from the traditional weighted average cost of capital formula found in the corporate finance literature. Now the firm's cash flow constraint (total use of funds must equal total sources) can be written as:

   (BC) $\quad RD_i + I_i + DIV = PRTS_i + DEP_i + NE_i + ND_i$, where in addition to the previously defined variables, we have

$PRTS_i \equiv$ current net profits for firm i

$DEP_i \equiv$ current depreciation and depletion for firm i.

As in Dhrymes-Kurz and McCabe, we treat current profits and depreciation expenditures as predetermined variables in the system, to be absorbed in the vectors of exogenous variables.

Using (BC) and equating the relevant marginal rates of returns to the modified marginal cost of financing equation, we can solve for RD_i, I_i, and DIV_i directly as follows:

(5) $RD_i = a_0 + a_1 ND_i + a_2 DIV_i + a_3 I_i + a_4 X_{rd_i}$,

with $a_0 = -\alpha_0/(\alpha_1 - \delta_0)$, $a_1 = (\delta_1 - \delta_0)/(\alpha_1 - \delta_0)$, $a_2 = \delta_0/(\alpha_1 - \delta_0)$,
$a_3 = (\delta_0 - \alpha_2)/(\alpha_1 - \delta_0)$, $a_4 = -\alpha_3/(\alpha_1 - \delta_0)$;

(6) $I_i = b_0 + b_1 ND_i + b_2 DIV_i + b_3 RD_i + b_4 X_{I_i}$,

with $b_0 = -\beta_0/(\beta_2 - \delta_0)$, $b_1 = (\delta_1 - \delta_0)/(\beta_2 - \delta_0)$, $b_2 = \delta_0/(\beta_2 - \delta_0)$,
$b_3 = (\delta_0 - \beta_1)/(\beta_2 - \delta_0)$, $b_4 = -\beta_3/(\beta_2 - \delta_0)$;

(7) $DIV_i = c_0 + c_1 ND_i + c_2 RD_i + c_3 I_i + c_4 X_{DIV_i}$,

with $c_0 = -\gamma_0/(\gamma_3 - \delta_0)$, $c_1 = (\delta_1 - \delta_0)/(\gamma_3 - \delta_0)$, $c_2 = (\delta_0 - \gamma_1)/(\gamma_3 - \delta_0)$,
$c_3 = (\delta_0 - \gamma_2)/(\gamma_3 - \delta_0)$, $c_4 = -\gamma_4/(\gamma_3 - \delta_0)$;

If we assume that firm's investment projects have different risk levels, we know that the cost of capital to the firm will be positively related to the firm's *BETA* or systematic riskiness, from the capital asset pricing model (see, for example, Copeland and Weston (1983, 399)). Hence, let us write the equation that represents the marginal cost of financing equation (5) as:

(8) $MCF_i = \zeta_0 + \zeta_1 BETA + \zeta_2 X_f$,

where X_f is a vector of exogenous variables and $\zeta_1 > 0$. Now combining (8), (4) and the firm's budget constraint, we can solve for the new debt variable as:

(9) $ND_i = d_0 + d_1 RD_i + d_2 DIV_i + d_3 I_i + d_4 BETA_i + d_5 X_{f_i}$,

where $d_0 = \zeta_0/\delta_1 - \delta_0$, $d_1 = -\delta_0/\delta_1 - \delta_0$, $d_2 = -\delta_0/\delta_1 - \delta_0$, $d_3 = -\delta_0/\delta_1 - \delta_0$,
$d_4 = \zeta_1/\delta_1 - \delta_0$, $d_5 = \zeta_2/\delta_1 - \delta_0$.

In deriving equation (9), we have chosen to eliminate new equity financing, through the balancing requirements of the budget constraint, following Dhrymes-Kurz and McCabe, due to a lack of reliable data on this variable. To the extent that new equity financing may be "nonnegligible" but still a "minor" source of external funds for the firms (compared to bond financing) as Dhrymes and Kurz remark (432–34), this "is not likely to lead to serious deficiencies." Although the data indicate that new equity financing in 1977 provided almost precisely the same share of total external financing as in the years pertinent to the Dhrymes and Kurz study, the reader should be cautioned that this (as well as Dhrymes-Kurz's and McCabe's) treatment of new equity as a balancing item can only be viewed as an approximation that may not be fully valid.

New equity issues (preferred and common stock) accounted for 25 percent of all new securities issued over the entire period 1963–77 and represented approximately 21 percent of all new issues in 1977, the year relevant to this study. See Survey of Current Business, *Business Statistics,* 1977 edition. The figure of 21 percent almost precisely obtains for two of the three years reported by Dhrymes and Kurz in their study (p. 433). One might thus assert that the precedence for treating new equity as a balancing item still obtains, although this should not be accepted without caution. For a discussion of the problems emanating from key "residual" items in financial models, see W. Brainard and J. Tobin, "Pitfalls in Financial Model Building," *American Economic Review* 58 (May, 1968), pp. 99–122.

At any rate, if we further assume that new debt can be issued at a risk free rate, we might conjecture that with the tax advantages of debt financing $(\delta_1 - \delta_0) < 0$.

J. Weston provides some evidence relating to the effects of the tax shield on debt which is consistent with a lower marginal cost of debt, as opposed to equity financing. See Weston, "A Test of Capital Propositions," *Southern Economic Journal* 30 (October, 1963), pp. 105–112.

Finally, in the model, short-term debt is excluded as a source of external financing for investment and dividends consistent with its treatment as a residual item by Dhrymes and Kurz (1967), McCabe (1979), and Nakamura and Nakamura (1982). (The latter assert (384–85) "the fact that long term debt is usually used to finance capital spending while short term debt is used to finance inventory and short term cash flow needs.")

7. See O. Willliamson, *Markets and Hierarchies: Analysis and Antitrust Implications* (New York: The Free Press, 1975), p. 179.

8. As Mueller notes (p. 73): "The faster a firm's sales are increasing, the more confidence it will have about its ability to secure the benefits from uncertain R & D and the more patience it can afford to show in waiting for these benefits. The faster a firm's sales are growing, the greater economic advantage it receives for a given cost reducing innovation." The problem with this variable, is that since it is possible that growth in sales may be correlated with past R & D (as in W. Leonard (1971, 1973)), what it may be capturing is merely the correlation between present and past R & D. Mueller remains sanguine, however, as he states: "If this is the case, the variable may serve as a measure of a firm's expectations. Firms continue to invest in R & D in the same relative proportions as they did in past, because their previous R & D outlays have been reinforced by the growth in their sales."

   It might be mentioned that J. Tilton's explanation for industry R & D allocations is thoroughly supportive of Mueller's interpretation. See J. Tilton, "Research and Development in Industrial Growth: A Comment," *Journal of Political Economy* 81 (September/October, 1973), pp. 1245–48.

9. E. Mansfield (1964, 32).

10. D. Jorgenson, "Econometric Studies of Investment Behavior: A Survey," *Journal of Economic Literature* 9 (December, 1971), pp. 111–47.

11. Ibid., p. 1111.

12. Ibid., p. 1112.

13. Ibid.

14. J. Lintner, "Distribution of Incomes of Corporations Among Dividends, Retained Earnings, and Taxes," *American Economic Review* 46 (May, 1956), pp. 97–113.

15. D. Mueller, p. 76; E. Kuh, *Capital Stock Growth: A Micro-Econometric Approach* (Amsterdam: North Holland, 1963), p. 17.

16. Note that Grabowski and Mueller actually measure this variable as well as the one that follows over eight year intervals.

17. G. McCabe, p. 134.

18. See, for example, Copeland and Weston, p. 441.

19. The primary data source used is the Standard and Poor's COMPUSTAT databank. Company betas were obtained from the Rodney White Center's publication, "Common Stock Risk Measures for the Period Ending June 30/77." Concentration ratios were constructed as averages from the concentration data in the U.S. 1972 *Census of Manufactures,* with the Standard and Poors

Corporate Directory providing four-digit SIC codes for the firms. Government R & D expenditures were obtained from Government Data Publication's *Research and Development Directory, 1977* ed.

20. Looking at total R & D (company plus government funded) as a percentage of sales, for our sample the computed expected value (based on a truncated lognormal distribution of R & D to sales ratios) is 0.031 which is the same as the total R & D to sales ratio of the NSF universe of manufacturing firms. For company funded R & D alone, in our sample, the analogous expected R & D to sales ration is 0.026, while the relevant ratio for the NSF universe is 0.020. With respect to the extent of coverage of our industries, the firms in our sample account for approximately 62% of all company funded R & D in the aerospace industry in 1977, 50.2% of all company funded R & D in the chemicals industry, 87.1% of all company funded R & D in the petroleum industry, and 44.9% of the relevant amount in the electronics industry. Finally, our firms as a whole accounted for approximately 24% of all company funded R & D in the United States in 1977.

21. The variables are defined as $D1$, $D2$, and $D3$. $D1$ equals one if the firm is classified into the petroleum industry (SIC 2911) and is zero otherwise; $D2$ equals one if the firm is classified into the electronics industry (SIC 3640, 3651, 3662, 3670, 3679, 3699) and is zero otherwise; $D3$ equals one if the firm is classified into the aerospace industry (SIC 3720, 3721, 3728, 3760) and is zero otherwise. Firms in the chemical industry embraced SIC 2800, 2810, 2860, 2870, and 2890.

22. In terms of the previous definitions, the variables that were deflated by sales (measured in millions of dollars) are RD, I, DIV, ND, IF, PRB, DEP, CF, CFT, $PRTS$, G, DS, $DS5$, DST, RDY, K, IY, and $DIVY$. Thus in examining Tables 1.1 and 1.2 the reader should be aware that the estimated coefficients for these variables refer to their deflated values, while the remaining variables maintain the definitions as set forth in the introduction.

23. To be more precise, internal financing relative to sales appears to be one of the central determinants of research and development intensity. The reader should take note that the exposition in the text assumes that the equations (ours as well as the models which are used for comparison purposes) are in level form—i.e., that both sides of each relevant equation are multiplied through by the company sales variable.

24. More direct evidence is provided by Mansfield (1984) and Mansfield and Switzer (1985). The latter study pertains to Canada, which unlike the U.S. has had a long history of experience with tax credits schemes for R & D. In this study, we found that although R & D tax credits do stimulate additional R & D, their effects have been modest. Furthermore, the foregone revenue to the government of the tax credits was found to exceed the stimulated R & D expenditure entailed.

25. The finding that R & D expenditures do not move immediately to desired levels but only about 30% of the way in a given year is consistent with an early work by Mansfield (1962).

   We should emphasize, however, that our estimates for 1977 are not fully comparable with the earlier ones, not simply because of the likely time dependence of the coefficient of lagged R & D, but also because our estimate is based on a larger and more heterogeneous sample of firms.

   The absence of complete adjustment is also consistent with a recent study of private R & D using macro data by Levy and Terleckyj (1983). On a priori grounds one might suggest that the inclusion of lagged R & D is essential.

   The partial adjustment process implied in the model (both for R & D and capital expenditures equations) may be overly simplified, though. Furthermore, the interpretation of the relevant lagged coefficients may not be straightforward. We have also estimated the system without the lagged R & D to sales ratio, with little change in the results, apart from a drop in significance in several of the coefficients and in the goodness of fit of the model. In particular the goodness of fit measure of Haessel (1978) falls in the R & D equation by 25% when lagged R & D to sales ratio is omitted. The corresponding declines in the capital expenditures and new debt equations are 3% and 2%, respectively, while the goodness of fit measure does not change for the dividend equation.

26. In making comparisons with other studies, it might be desirable to account for possible effects of the business cycle. The studies that might be compared (as well as contrasted) to ours utilize data from cross sections, time series and pooled cross section-time series for various years over the period 1951–76. Our study relates to but one year, 1977, which might be deemed an "upswing" year. It was not always possible to obtain cross sectional estimates of the various coefficients corresponding to "upswing" years or periods. Hence, aside from systematic differences between our results and the results of others due, for example, to structural change, there may be cyclical influences that bias certain comparisons that we make. For three of the main cross section studies that overlap with ours, Dhrymes and Kurz (1967), Mueller (1967) and McCabe (1979), however, we were able to select the years 1951, 1959, and 1972 as "upswing" years, and thus when comparisons are made between certain coefficients we estimated for 1977 and their counterparts in Dhrymes and Kurz, Mueller and McCabe, we will focus our comparisons with the results for 1951, 1959, and 1972.

27. The 3SLS procedure outperformed 2SLS and OLS estimation in terms of generating coefficients that are statistically significant at $\alpha = 0.05$ or better (15 as opposed to 14). Although the OLS, 2SLS, and 3SLS results are largely similar in terms of sign, magnitude and significance of estimated coefficients, there are some important differences that should be mentioned. For the R & D equation, moving from OLS to 2SLS to 3SLS we found that the effect of internal funding declines in magnitude and significance, as might be expected. In the capital expenditures equation, while new debt is positive and significant in the OLS estimation, it loses significance in 2SLS and 3SLS estimation. Lagged capital stock is not significant at all in 2SLS, unlike the 3SLS result.

In the dividend equation, while investment is negative and significant in the simultaneous procedures, it is not significant in OLS. Further, new debt is positive and significant in 3SLS (as might be expected in a funds flow perspective); it is not significant in OLS and 2SLS. Slight statistical significance of the NYSE dummy is observed in OLS, which disappears in 2SLS and 3SLS.

In the new debt equation, while current cash flow (*CFT*) is mildly significant and of expected sign in 2SLS and 3SLS, it lacks significance in the OLS estimation.

Overall, these results illustrate the potential problems of the OLS single equation and the inefficient 2SLS simultaneous approaches.

Chapter 2

1. Morton I. Kamien and Nancy L. Schwartz, "Market Structure and Innovation: A Survey," *Journal of Economic Literature* 13 (1975), p. 2.

2. E. Mansfield, *The Economics of Technological Change* (New York: Norton, 1968), p. 186.

3. Ibid.

4. *Economic Report of the President* (Washington, 1972), p. 126.

5. Mansfield, 1968, p. 187.

6. J. Hirshleifer, on the other hand, developed various models of technological uncertainty which he feels bring into question the conclusion that the private sector lacks sufficient incentives to

generate a socially optimal level of inventive activity. In the presence of heterogeneous beliefs, Hirshleifer shows that conflicting expectations of distributive gain (in the form of differing expectations regarding wealth transfers caused by price revaluations of private goods whose production is related to the technological information yielded by research) will create excessive incentives (in which social costs exceed social benefits) even for the generation of public information in a simple model of pure exchange. In the more realistic setting of production and exchange, though, he shows that there may or may not be overinvestment in information generating activities (such as R & D aimed at discovery of natural phenomena) depending on the extent of gains resulting from rearrangement of productive activities relative to the costs of acquiring and disseminating the information. One might remark that the case for government support of R & D, especially in areas where potential social gains are great (as opposed to the "horse race" paradigm, where of course the social value of information may not be great), is not at all ruled out by this latter model of production and exchange. For as he remarks, "there is no logically necessary tie between the size of the technological benefit on the one hand, and the amplitude of the price shift, that create speculative opportunities." The case for government support of private R & D is further strengthened by noting that the models do not account for the fact that highly imperfect markets for the time-state claims may prevent innovator-speculators from deriving any monetary gains that accompany preferred access to information which once released becomes a public good. See Hirshleifer, "The Private and Social Value of Information and the Reward to Innovative Activity," *American Economic Review* 61 (September, 1971), pp. 561–74.

7. D. Blank and G. Stigler, *The Demand and Supply of Scientific Personnel* (New York: NBER, 1957), p. 57.

8. Ibid., pp. 57–58.

9. Ibid., p. 58.

10. Ibid.

11. Ibid., p. 59.

12. Ibid., p. 62.

13. W. Leonard, "Research and Development in Industrial Growth," *Journal of Political Economy* 79 (March/April, 1971), p. 245.

14. Ibid., p. 250.

15. Ibid.

16. O. Williamson, *Markets and Hierarchies: Analysis and Antitrust Implications* (New York: The Free Press, 1975), p. 191.

17. N. Terleckyj, "Direct and Indirect Effects of R & D on the Productivity Growth of Industry," in J. Kendrick and B. Vaccara, eds., *New Developments in Productivity Measurement and Analysis* (Chicago: University of Chicago Press, 1980), pp. 359–77.

18. S. Globerman, "Comment," in Kendrick and Vaccara eds., op. cit., p. 383.

19. Z. Griliches, "Returns to Research and Development in the Private Sector," in Kendrick and Vaccara, eds., op. cit., p. 445.

20. Ibid., pp. 445–46.

21. The definition of energy-related research and development we use is the NSF definition and encompasses "all R & D spending whose purpose is to increase energy resources or capabilities,

including the development of energy equipment. Energy R & D can include costs of R & D projects (both product and process) on exploration, extraction, transportation, processing, storage, generation (including conservation, etc.) of present, new, or improved forms of energy.''

22. *Business Week,* June 27, 1977, p. 52.

23. National Science Foundation, *Federal R & D Funding by Budget Function: Fiscal Years 1979–81,* p. 28.

24. R. W. Schmitt and P. J. Stewart, ''Unrealities in the Energy R & D Program,'' *Research Management* 21 (July, 1973), p. 7.

25. Ibid., p. 9.

26. Ibid.

27. *Business Week,* June 27, 1977, p. 55.

28. U.S. Government, *Code of Federal Regulations,* Title 41, Public Contracts and Property Management (Washington: U.S. Government Printing Office, 1979), pp. 81–82.

29. Ibid.

30. It might be noted that H. Armour and D. Teece also show that this holds for earlier periods as well for the petroleum industry. In addition, they demonstrate that for the petroleum industry over the period 1959–76, the top 4, top 8, and top 20 firms accounted for 45, 70, and 87 percent of all innovations, 51, 67, and 98 percent of all process innovations, and yet only 35, 61, and 90 percent of the respective refining capacity in the industry. These results are used to bolster their argument against proposed measures at divestiture of the assets of firms in this industry. See Teece and Armour, ''Innovation and Divestiture in the U.S. Oil Industry,'' in D. Teece, ed., *R & D in Energy: Implications of Petroleum Industry Reorganization* (Palo Alto: Institute for Energy Studies, 1977), pp. 7–97.

31. E. Roberts, *The Dynamics of Research and Development* (New York: Harper and Row, 1964), p. 83.

32. Ibid.

33. Ibid.

34. O. Williamson, ''Transaction Cost Economics: The Governance of Contractual Relations,'' *Journal of Law and Economics* 22 (October, 1979), pp. 233–61.

35. This is also apparent in the sample of firms which we study in this chapter. For these firms, the simple correlation coefficient between R & D expenditures and sales is significantly greater than zero at the 0.05 level.

36. Z. Griliches, 1979, p. 106.

37. This is based on the figures in the NSF publication, *Research and Development in Industry, 1978* (Washington, D.C.: U.S. Government Printing Office, 1979).

38. The relevant population figures here are obtained from Business Week's Survey of Corporate Research and Development Spending for 1976; these figures are in turn based on COMPUSTAT tabulations, which are disaggregated by four-digit SIC class, as opposed to the two-digit and three-digit SIC breakdowns used by NSF.

39. A copy of the questionnaire is available from the author on request.

40. Federal regulations in regard to waiving the government's rights for DOE sponsored contracts require that in making waiver determinations,

The Head of the Agency or designee shall have the following objectives:

(1) Making the benefits of the energy research, development, and demonstration program widely available to the public in the shortest possible time;
(2) Promoting the commercial utilization of such inventions;
(3) Encouraging participation by private persons in DOE's energy research, development, and demonstration program; and
(4) Fostering competition and preventing undue market concentration or the creation or maintenance of other situations inconsistent with the antitrust laws.

See U.S. Government, Code of Federal Regulations, Title 41, *Public Contracts and Property Management,* Chapter 9, The Department of Energy (Revised as of July 1979) (Washington, D.C.: U.S. Government Printing Office, 1979), pp. 256–67.

41. See for example E. Mansfield, ''Basic Research and Productivity Increase in Manufacturing,'' *American Economic Review* 70 (December 1980), pp. 869–70.

42. This is the weighted average, based on 39 projects with weights proportional to the average annual expenditures on the projects. Using total expenditures on the projects as the relevant weights, this averge rises to 0.715. The unweighted mean of this variable is 0.596.

43. In only 3 out of 41 cases would the firms have carried out all of the work without government support.

44. This is the weighted average for 41 projects, with weights assigned according to the average annual expenditures on the projects. Using total expenditures on the projects as the relevant weights, this average falls to 15.8 percent. The unweighted mean of this variable is 0.203.

45. See for example, G. Maddala, *Econometrics* (New York: McGraw-Hill, 1977), p. 164.

46. An alternative derivation would be to assume that the ex ante probability that $Y_n = 1$ is exactly given by

$$P_n = 1/1 + e^{-X_n\beta},$$

and for $Y = 0$, the ex ante probability is

$$(1 - P_n) = 1/1 + e^{+X_n\beta}.$$

If we wished to modify this aproach to allow for the possibility that actual probabilities may deviate randomly around some expected value, e.g., let

$$P_n = 1/1 + e^{-X_n\beta + V_n}.$$

Where V_n is the disturbance term, the problem becomes computationally intractable, however. See E. Hanushek and J. Jackson, *Statistical Methods for Social Scientists* (New York: Academic Press, 1977), p. 203.

47. See, for example, H. Theil, *Principles of Econometrics* (New York: Wiley, 1971), pp. 384–85.

48. See E. Hanushek and J. Jackson, op. cit.

49. To be precise, letting $\hat{L}1 = \ln L1$, at the maximum we have the first order conditions:

$$\partial\hat{L}1/\partial\hat{\beta}_k = \Sigma_{n=1}^{N} Y_n X_{nk} - \Sigma_{n=1}^{N} X_{nk}/1 + e^{-X_n\beta} = 0 \text{ for } k = 1, \ldots, K.$$

50. The procedure is described in G. Maddala, op. cit., pp. 173–75. The criterion we employed for convergence was that the calculated change in the gradient be less than 0.0000001 for two straight iterations.

51. With regard to institutional factors, our concern is with government sponsored energy R & D that is conducted in private R & D labs; the performance of quasi-public Federally financed R & D centers that are managed by the private sector (e.g., the administration of the Oak Ridge National Laboratory by Union Carbide) is not examined here.

   With regard to political factors, one might remark that an incident such as occurred at Three Mile Island would, for example, be expected to influence the private sector's willingness to perform R & D aimed at nuclear energy. It may be of interest to note that in our sample, two of the projects were concerned with nuclear energy, one which developed an idea that originated within the firm, and the other which was suggested solely by the government. They were both conducted in periods that overlapped with the TMI reactor accident. In only one case (where the idea for the project originated within the firm) would the firm have supported the project in the absence of government financing (the amount of private funding that would have been forthcoming would have supported only 50 percent of the work done on the project, however). We also note that the project that would not have been abandoned entirely in the absence of government funding led to additional R & D projects into which the firm invested its own funds.

52. For example, dimensionalizing projects according to the ex ante percentage contribution to the firm's technological objectives and the extent of separation, for 36 projects we get the following table:

Expected Dollar Value of Project's Contribution to Firm's Technological Objectives Compared to One Dollar Spent on Non-Federally Funded Energy R & D was ≤ $0.50	Project's Resources Kept Completely Separate from Private R & D Projects	Project's Resources Not Completely Separated from Private R & D Projects	Total Number of Cases
	5	10	15
Expected Dollar Value was > $0.50	1	20	21
Total Number of Cases	6	30	36

The χ^2 statistic associated with this table is 6.75, which is significantly greater than the critical value (with one degree of freedom) for a 5 percent level of significance of 3.84. We must reject the null hypothesis of a random association between contribution to technological objectives and separation, based on this table. Alternatively, specifying a logit model explaining this separation dichotomy, we find that the odds that a project's resources are kept separate is positively and highly significantly related to the expected dollar value of the project's contribution to the firm's technological objectives compared to one dollar spent on non-Federally funded energy R & D (measured as a continuous variable). This also holds for an alternative dichotomization of the separation variable as well as for the ex post measure of "contribution to technological objectives."

53. As in the note above, classification of the projects according to the (ex ante) percent contribution of the project to the firm's overall technological objectives and the firm involvement/noninvolvement in the generation of the ideas for the projects, we get the following contingency table:

Expected Dollar Value of Project's Contribution to Firm's Technological Objectives Compared to One Dollar Spent on Non-Federally Funded Energy R & D was ≤ $0.20	Firm was Involved in the Generation of the Idea for the Project	Idea for Project was not Attributable to Firm	Total Number of Cases
	7	8	15
Expected Dollar Value was > $0.20	3	18	21
Total Number of Cases	10	26	36

The χ^2 statistic computed from this table is 4.58, which again is statistically significant at the 5 percent level. Specifying a logit model explaining the source of idea dichotomy, it was found that the odds that the project's idea originated outside the firm is positively and significantly related to the contribution of the project to the firm's technological objectives (either ex ante or ex post) as would be intuitively expected.

54. The calculations of the *t*-values for the maximum likelihood logit models proceeded as follows. Under general conditions, MLEs are consistent and asymptotically normal, with an asymptotic variance-covariance matrix given by the inverse of the information matrix.

Hence, writing the log of the likelihood function as $L1$, the variance-covariance matrix as *VCOV*, we have

(1) $VCOV = -E\{\partial^2 \hat{L}1/\partial\beta\partial\beta'\}^{-1}$,

where E is the expectations operator; for the logit model we estimated, this simplifies to

(2) $VCOV = \{\sum_{i=1}^{N} [(e^{X_i'\beta}/(1 + e^{X_i'\beta}) (1/1 + e^{X_i'\beta}) X_i X_i']\}^{-1}$,

where X_i is the K element of explanatory variables for the ith observation and N is the total number of observations.

Substituting the MLE parameter estimates $\hat{\beta}_{MLE}$ into *VCOV*, we obtained our estimate of the variance-covariance matrix. Coefficient standard errors were then obtained by taking the square roots of the diagonal of *VCOV*. The individual coefficients $\hat{\beta}_k$, $k = 1, \ldots, K$ divided by their standard errors are the *t*-values that are shown in the text and are asymptotically distributed as t statistics with N-K degrees of freedom (which in turn correspond asymptotically to the standard normal distribution). See, e.g., T. Amemiya, "Qualitative Response Models: A Survey," *Journal of Economic Literature* 19 (Dec., 1981), pp. 1483–1536.

Chapter 3

1. F. A. Hayek, "The Use of Knowledge in Society," *American Economic Review* 35 (September, 1945), pp. 526–27.

2. From the firm's point of view, high rates of inflation in R & D relative to other activities might serve as a stimulant or retardant to nominal investment in R & D. If firms wish to maintain a given level of research effort over time, nominal expenditures might be expected to rise with inflation pari passu. To the extent that high rates of inflation adversely affect the firm's cost of capital or increase the attractiveness of alternative uses of the firm's investment funds (such

as investment in securities), relative allocations towards R & D may be expected to decline, on the other hand. Regarding the latter possibility, in Mansfield's recent study, for example, one of the reasons adduced for the cutting back on the proportion of R & D expenditures aimed at basic research and relatively long term projects was inflation. See Mansfield, "Basic Research and Productivity Increase in Manufacturing," *AER* 70 (December, 1980), p. 871.

3. The annual costs included salaries, equipment and operating expenses, and rent, and excluded capital investments (with the exception of depreciation charges. See H. Milton, "Cost of Research Index, 1920–1965," *Operations Research* 14 (November/December, 1966), p. 978, and "Cost of Research Index, 1920–1970," *Operations Research* 20 (January/February, 1972), p. 2.

4. H. Milton (1972), p. 1.

5. Ibid., p. 16.

6. Z. Griliches, "Issues in Assessing the Contribution of Research to Productivity Growth," *Bell Journal of Economics* 10 (Spring, 1979), p. 105.

7. S. Jaffe, "A Price Index for Deflation of Academic R & D Expenditures," NSF #72-310, May 1972, p. 3.

8. Ibid., p. 2.

9. Again, basic research, according to the NSF definition, encompasses "original investigations for the advancement of scientific knowledge not having specific commercial objectives, although such investigations may be in fields of present or potential interest to the reporting company." Applied research covers "investigations directed to the discovery of new scientific knowledge having specific commercial objectives with respect to products or processes. This definition differs from that of basic research chiefly in terms of the objectives of the reporting company." Development is defined as "technical activities of a nonroutine nature concerned with translating research findings or other scientific knowledge into products or processes. [It] does not include routine technical services to customers, quality control, routine product testing, market research, sales promotion, sales service, research in the social sciences or psychology, and other nontechnological activities or technical services." See NSF, *Research and Development in Industry,* 1978.

   Given these definitions, in 1979, for example, it is estimated that approximately 78 percent of industry's R & D performance was for development, 19 percent for applied research, and 3 percent for basic research. For universities, the corresponding figures are 70 percent for basic research, 25 percent for applied research, and 5 percent for development. See *Chemical and Engineering News,* July 28, 1980. In 1967 (the year relevant to Jaffe's base-weight calculations), the breakdown for industry is 3.8 percent for basic research, 17.8 percent for applied research and 78.4 percent for development. For universities and colleges in 1967, the distribution is 77.4 percent for basic research, 18.9 percent for applied research, and 3.7 percent for development. See for example NSF, *Research and Development in Industry,* 1968, and NSF, *Resources for Scientific Activities at Universities and Colleges,* 1971. Thus there may be some difficulties associated with the use of this index for industrial R & D given that the nature of most R & D is not categorized as basic research, but rather as development.

10. It might be mentioned that NASA has also developed a price index for its R & D (which has increased at a somewhat faster rate than the GNP deflator over the period 1972–79). However, extrapolating this index to industrial R & D may be unwarranted for the same reasons as adduced for the Jaffe index.

11. See for example Mansfield et al., *The Production and Application of New Industrial Technology* (New York: W. W. Norton, 1977, chapter 3).

12. F. Fisher and K. Shell, *The Economic Theory of Price Indices* (New York: Academic Press, 1972), p. 3.

13. See, for example, W. Diewert, "The Economic Theory of Index Numbers: A Survey," U.B.C. Discussion Paper #79-09, March, 1979, p. 44.

14. This can be shown as follows. The fixed coefficient form can be written as

$$y = \min \{X_1/a_1, X_2/a_2, \ldots, X_n/a_n\}.$$

At the optimum, for output level y^0 corresponding to the base period with an input vector X^0, we have

$$y^0 = X_i^0/a_i \text{ or alternatively } a_i y^0 = X_i^0 \; \forall_i = 1, \ldots, n.$$

$$\Rightarrow C^*(W^0, y^0) = \Sigma_{i=1}^{n} W_i^0 a_i y^0 = \Sigma_{i=1}^{n} W_i^0 X_i^0.$$

Similarly, $C^*(W^1, X^0) = \Sigma_{i=1}^{n} W_i^1 a_i y^0 = \Sigma_{i=1}^{n} W_i^1 X_i^0.$

$$\Rightarrow P^* = \Sigma_i W_i^1 X_i^0 / \Sigma_i W_i^0 X_i^0,$$

which is identical to the Laspeyres index. Using current output level y^1 as the reference iso-quant, the argument in the text regarding the Paasche index follows analogously.

15. Recall that a homothetic production function is of the form $F(\Phi(X))$ where F is monotonically nondecreasing and $\Phi(X)$ is homogeneous of degree one. Given such a production function, marginal rates of substitution among inputs are constant along any ray emanating from the origin. The proposition in the text follows from the factorization property of cost functions corresponding to homothetic production functions. See, for example, R. Shephard, *Theory of Cost and Production Functions* (Princeton: Princeton University Press, 1970), p. 93.

16. This argument extends the Fisher and Shell propositions regarding the impact on the CPI of taste changes. See Fisher and Shell, op. cit., p. 6.

17. Organization for Economic Cooperation and Development, *Trends in Industrial R & D in Selected OECD Member Countries, 1967-75* (Paris: OECD, 1979), p. 158.

18. Representative, that is, in terms of his perception of their technological contributions as well as their commercial potential.

19. The R & D to sales ratio for agricultural chemicals is higher than for any other line of business for the company. One of the reasons adduced for this is that the cost of hydrocarbons, which has been a major element of the nation's inflation since 1973, only accounts for a minor fraction of the final selling price of agricultural chemicals.

20. The stages of product innovation in this framework are sixfold, and are defined as follows:

    (a) Applied research—investigations directed to the discovery of new scientific knowledge having specific commercial objectives, with respect to products or processes. This definition differs from that of basic research chiefly in terms of the objectives of the company.
    (b) Preparation of project requirements and basic specifications—routine planning and scheduling which involves close coordination with marketing to increase the probability that the project will be a commercial success. Bench-scale work and applications research would be included in this activity.
    (c) Pilot plant and prototype design and construction—this would also include any work related to the manufacturing of the product in quantity.
    (d) Tooling and construction of manufacturing facilities—this activity focuses on the problems of manufacturing and may involve the transfer of responsibility from R & D management to production management.

(e) Manufacturing startup—this would involve the training of workers, "debugging" the plant, and any production before an acceptable quality level is reached.

(f) Marketing startup—this would include market studies, advertising, the establishment of a system of distribution, and other sales-related expenditures incurred before the first sale and delivery of the product.

Note that in this framework, only the first three of these stages would be encompassed by the NSF definition of R & D. See Mansfield et al., *The Production and Application of New Industrial Technology,* and Mansfield et al., *Research and Innovation in the Modern Corporation* (New York: W. W. Norton, 1971).

21. It was claimed that although the actual length of time required to complete the registration process varies, on average in the industry it takes six to seven years for a firm to complete the registration tests and 18 to 24 months for the EPA to review the data. The increased stringency of the EPA's registration requirements was recognized by corporate officials as one of the key contributing factors to the recent inflation in costs of R & D in agricultural chemicals as well as perceptions of expected future inflation. For example, one of the senior scientists whom we interviewed from the company presented us with a study he completed on the impact of new Federal guidelines (as proposed in the Federal Register of July 10, 1973) on analytical research. These guidelines require that the firm have available analytical methods for detecting and measuring the concentration of all ingredients and impurities that are present in quantities of 0.01 percent or more of the weight of the product. Actual measurements to this level for any given compound are not necessarily mandatory, however. Using as a benchmark a product in which the firm attempted to identify all compounds present at the 0.1 percent level or more (done at a cost of 5 percent of the total research cost for the product—which was a miniscule part of the firm's R & D program), he estimated that for the product alone, expenditures on the order of one-half or more of the firm's entire R & D expenditures would be required to identify all compounds at the 0.01 percent level. Given this evidence, he asserted that the imposition of mandatory requirements on identification of all ingredients for all compounds (a direction which the proposals of 1978 may lead) will entail costs way out of proportion to potential benefits. His impression is that the need for Federal regulation of toxic chemicals to protect consumers and plant workers is satisfied at present by the EPA's right to ask for analyses below 0.01 percent for highly toxic compounds; further restrictions, however, are seen as detrimental to the industry's survival.

22. Other direct and outside services which were of minor importance (accounting for about 10% of the expenditures in 1969) comprised six subcategories: (a) travel, (b) depreciation, (c) utilities, (d) other direct charges—which comprise taxes, licenses, insurance fees, library costs, seminars, postage and freight, and rentals, (e) external engineering consulting, and (f) building renovations. The relevant price ratios are shown below:

Travel	Depreciation	Utilities	Other Direct Charges	External Engineering Consulting	Building Renovations
99.3	90	196	365	216	223

23. Educational attainment groupings were according to university degree attained, (B.Sc., M.Sc., Ph.D.). Academic discipline groupings identified included chemists, physicists, mathematicians, and various engineering subgroups. Experience categories differentiated individuals with less than three years experience, three to five years, and over five years of experience.

24. The costs represented by item (f) are the same as those described in n. 22 above.

25. It might be mentioned that the substantial rate of increase in the costs of capital equipment services and of analytical equipment may be largely attributable to the increased stringency of the

Environmental Protection Agency's registration requirements and the increased sophistication of capital and analytical equipment that need be employed to meet EPA standards. This factor was alluded to by one of the senior scientists of the corporation who assisted in the data collection.

26.   In these cases, firms usually provided proxies, such as personnel remuneration and/or total expenditure series, but these were of little use for our purposes.

27.   The firms in the sample account for approximately 45 percent of the R & D expenditures in the petroleum industry, 30 percent of the R & D expenditures in the primary metals industry, 20 percent of the R & D expenditures in the chemical industry, 12 percent of the R & D expenditures in the rubber industry, 10 percent of the R & D expenditures in the electrical equipment industry, 7 percent of the R & D expenditures in the fabricated metal products industry, and less than 5 percent of the R & D expenditures in the stone, clay, and glass and textile industries. Of course the industries do not exhaust the entire range of industries in the NSF universe, but as is noted in the text they represent a substantial proportion of company funded R & D in the U.S.

28.   Due to data limitations, R & D expenditures of 1976, rather than 1969, were used as weights. As the available data indicates that R & D expenditures in 1976 are proportional to expenditures in 1969, the indices are essentially equivalent to Laspeyres indices.

29.   L. Goldberg, "Federal Policies Affecting Industrial Research and Development," presented at the meeting of the Southern Economic Association, November 9, 1978.

30.   National Science Foundation, *Research and Development in Industry, 1970* (Washington, D.C.: U.S. Government Printing Office, 1972).

31.   Organization for Economic Cooperation and Development, op. cit., p. 6.

32.   See for example Mansfield, "Technology and Productivity in the United States: Developments and Changes in the Postwar Period," in M. Feldstein ed., *The American Economy in Transition* (Chicago: NBER, 1980); and National Science Foundation, *Science Indicators* (Washington, D.C.: U.S. Government Printing Office, 1979).

33.   Note that the Laspeyres and Cobb-Douglas indices in Table 3.3 are not fully compatible, however, for reasons alluded to earlier in the text.

34.   It might be mentioned that only the NSF estimates of α_i are consistent with an unchanging homothetic Cobb-Douglas technology for chemicals (with no allowance necessary for the effects of regulatory policies and other factors that may have affected the relative as well as absolute productivity of the various R & D inputs) since only in this case does the index lie between the Laspeyres and Paasche bounds. Unfortunately, we lacked the necessary data to estimate an alternative technology structure and/or to account explicitly for such countervailing forces as gains in instrumentation and increased stringency of the regulatory policies of the EPA and OSHA.

35.   That is, using the usual *t*-test with a critical level of 5 percent assumed.

36.   Mansfield et al., 1977, p. 69.

37.   Ibid., p. 72.

38.   As complete data on all the firms' innovation costs were not available, we assumed that each firm's innovation costs are proportional to 1976 R & D expenditures, which accords with the information available.

39.   With regard to the tooling and construction stage, one firm estimated an increase in price of 125 percent *after* accounting explicitly for the effects of government regulations and controls, *as well as* the increased efficiency brought about by improvements in technology.

40. Battelle Memorial Institute, *Probable Levels of R and D Expenditures in 1980: Forecast and Analysis* (Columbus, Ohio: Battelle, 1979).

Chapter 4

1. E. Mansfield et al., *The Production and Application of New Industrial Technology* (New York: W. W. Norton & Co., 1977), p. 195.

2. Ibid., p. 194.

Bibliography

Amemiya, T. "Qualitative Response Models: A Survey." *Journal of Economic Literature* 19 (December, 1981), pp. 1483–1536.

Arditti, F. D. "Risk and the Required Return on Equity," Ph.D. Dissertation, Massachusetts Institute of Technology, 1965.

Battelle Memorial Institute. *National Survey of Compensation: Paid Scientists and Engineers Engaged in Research and Development Activities, 1969.* Columbus, Ohio: Battelle Columbus Laboratories, 1969.

_____. *National Survey of Compensation: Paid Scientists and Engineers Engaged in Research and Development Activities, 1979.* Columbus, Ohio: Battelle Columbus Laboratories, 1979.

_____. *Probable Levels of R and D Expenditures in 1980: Forecast and Analysis.* Columbus, Ohio: Battelle, 1979.

Black, G. *Substitution of Public for Private Research and Development Expenditures.* Organization Research Program Working Paper #57-64, Sloan School of Management, MIT, 1964.

_____ and W. A. Fischer. "Federal Funding of Industrial R & D: Stimulus or Substitute?" *Research Management* 20 (1979), pp. 27–30.

Blank, D. M., and G. J. Stigler. *The Demand and Supply of Scientific Personnel.* New York: NBER, 1957.

Branch, B. "Research and Development and Profitability: A Distributed Lag Analysis." *Journal of Political Economy* 82 (1974), pp. 999–1011.

Brown, M., and A. Conrad. "The Influence of Research on CES Production Relations." In M. Brown, ed. *The Theory and Empirical Analysis of Production.* New York: Columbia University Press for NBER, 1967, pp. 275–340.

Business Week, July 6, 1981.

Carmichael, J. "The Effects of Mission Oriented Public R & D Spending on Private Industry." *Journal of Finance* 36 (1981), pp. 617–27.

Chemical and Engineering News, July 13, 1981.

Christainsen, G., and R. Haveman. "Public Regulations and the Slowdown in Productivity Growth." *American Economic Association, Papers and Proceedings* 71 (1981), pp. 320–25.

Comanor, W. "Market Structure, Product Differentiation, and Industrial Research." *Quarterly Journal of Economics* 31 (1967), pp. 639–57.

Copeland, T., and J. Weston. *Financial Theory and Corporate Policy.* Reading, Mass.: Addison-Wesley, 2nd edition, 1983.

Dhrymes, P., and M. Kurz. "Investment, Dividends, and External Finance Behavior of Firms." In R. Ferber, ed., *Determinants of Investment Behavior.* New York: Columbia University Press, 1967, pp. 427–67.

Diewert, W. "The Economic Theory of Index Numbers: A Survey." U.B.C. Discussion Paper #79-09, March 1979.

Fama, Eugene. "The Empirical Relationships between the Dividend and Investment Decision of Firms." *American Economic Review* 64 (1974), pp. 304-18.

Fisher, F., and K. Shell. *The Economic Theory of Price Indices.* New York: W. W. Norton, 1972.

Fortune. *Plant and Product Directory,* 1967 edition.

The Globe and Mail. June 2, 1980, p. B8.

Globerman, S., "Market Structure and R & D in Canadian Manufacturing Industries." *Quarterly Review of Economics and Business* 13 (1973), pp. 59-68.

———. "Comment." In J. Kendrick and B. Vaccara, eds., *New Developments in Productivity Measurement and Analysis.* Chicago: University of Chicago Press, 1980, p. 383.

Goldberg, L. "Federal Policies Affecting Industrial Research and Development." Paper presented at the meetings of the Southern Economic Association, November 9, 1978.

Grabowski, H. "The Determinants of Industrial Research and Development: A Study of the Chemical, Drug, and Petroleum Industries. *Journal of Political Economy* 76 (1968), pp. 292-306.

———, and D. Mueller. "Managerial and Stockholder Models of Firm Expenditures." *Review of Economics and Statistics* 54 (1972), pp. 9-24.

Griliches, Zvi. "Research Expenditures, Education and the Aggregate Agricultural Production Function." *American Economic Review* 54 (December, 1964), pp. 961-74.

———. "Issues in Assessing the Contribution of Research and Development to Productivity Growth." *Bell Journal of Economics* 10 (1979), pp. 92-116.

———. "Returns to Research and Development in the Private Sector." In J. W. Kendrick and B. Vaccara, eds., *New Developments in Productivity Measurement and Analysis.* Chicago: University of Chicago Press, 1980, pp. 419-54.

Haessel, Walter. "Measuring Goodness of Fit in Linear and Nonlinear Models." *Southern Economic Journal* 44 (1978), pp. 648-52.

Haley, C., and L. Schall. *The Theory of Financial Decisions.* New York: McGraw-Hill Inc., 1979.

Hamberg, D. *R & D: Essays on the Economics of Research and Development.* New York: Random House, 1966.

Hanushek, E., and J. Jackson. *Statistical Methods for Social Scientists.* New York: Academic Press, 1977.

Hayek, F. A. "The Use of Knowledge in Society." *American Economic Review* 35 (1945), pp. 519-30.

Hellwig, M. "Bankruptcy, Limited Liability, and the Modigliani-Miller Theorem." *American Economic Review* 71 (March, 1981), pp. 155-70.

Hirshleifer, J. "The Private and Social Value of Information and the Reward to Innovative Activity." *American Economic Review* 61 (1971), pp. 561-74.

Horwitz, B., and R. Kolodny. "The FASB, the SEC, and R & D." *Bell Journal of Economics* 12 (Spring, 1981), pp. 249-62.

Howe, J. D., and D. G. McFetridge. "The Determinants of R & D Expenditures." *Canadian Journal of Economics* 9 (1976), pp. 57-71.

Jaffe, S. "A Price Index for Deflation of Academic R & D Expenditures." NSF #72-310, May 1972.

Jorgenson, D. "Econometric Studies of Investment Behavior: A Survey." *Journal of Economic Literature* 9 (Dec., 1971), pp. 111-47.

Kamien, M. I., and N. L. Schwartz. "Market Structure and Innovation: A Survey." *Journal of Economic Literature* 13 (1975), pp. 1-37.

———. "Self-Financing of an R & D Project." *American Economic Review* 68 (June, 1978), pp. 252-61.

Kuh, E. *Capital Stock Growth: A Micro-Econometric Approach.* Amsterdam: North-Holland, 1963.

——— and J. Meyer. "Correlation and Regression Estimates When the Data are Ratios." *Econometrica* 23 (October, 1955), pp. 400-15.

Leonard, W. N. "Research and Development in Industrial Growth." *Journal of Political Economy* 79 (1971), pp. 232-56.

_____. "Reply." *Journal of Political Economy* 81 (1973), pp. 1249–52.

Levin, R. "Toward an Empirical Model of Schumpeterian Competition." NBER Summer Institute Paper #80-11, 1980.

Levy, David M., and Nestor E. Terleckyj. "Effects of Government R & D on Private R & D Investment and Productivity: A Macrocosmic Analysis," *Bell Journal of Economics* 14 (1983), pp. 551–61.

Link, A. "A Disaggregated Analysis of R & D Spending." *Southern Economic Journal* 49 (Oct., 1982), pp. 342–49.

Lintner, J. "Distribution of Incomes of Corporations Among Dividends, Retained Earnings, and Taxes. *American Economic Review* 46 (1956), pp. 97–113.

Maddala, G. *Econometrics.* New York: McGraw-Hill, 1977.

Mansfield, E. "Comment." In *The Rate and Direction of Inventive Activity: Economic and Social Factors.* Princeton, N.J.: Princeton Univ. Press, 1962, pp. 188–94.

_____. "Industrial Research and Development Expenditures: Determinants, Prospects, and Relation to Size of Firm and Inventive Output." *Journal of Political Economy* 62 (1964), pp. 319–40.

_____. "Rates of Return from Industrial Research and Development." *American Economic Review* 55 (May, 1965), pp. 310–22.

_____. *Industrial Research and Technological Innovation—An Econometric Analysis.* New York: W. W. Norton for the Cowles Foundation for Research in Economics at Yale University, 1968a.

_____. *The Economics of Technological Change.* New York: W. W. Norton, 1968b.

_____. "Basic Research and Productivity Increase in Manufacturing." *American Economic Review* 70 (1980), pp. 863–73.

_____. "Technology and Productivity in the United States: Developments and Changes in the Postwar Period." In M. Feldstein, ed., *The American Economy in Transition.* Chicago: NBER, 1980.

_____. "How Effective is the R & D Tax Credit," *Challenge* 27 (1984), pp. 57–61.

_____, J. Rapoport, J. Schnee, S. Wagner, and M. Hamburger. *Research and Innovation in the Modern Corporation.* New York: W. W. Norton, 1971.

_____, J. Rapoport, A. Romeo, E. Villani, S. Wagner, and F. Husic. *The Production and Application of New Industrial Technology.* New York: W. W. Norton, 1977.

_____, and L. Switzer. "The Effects of R & D Tax Credits and Allowances in Canada." *Research Policy* 12 (1985), pp. 97–107.

_____, A. Romeo, and L. Switzer. "R & D Indexes and Real R & D Expenditures." *Research Policy* 12 (1983), pp. 105–12.

McCabe, G. "The Empirical Relationship Between Investment and Financing: A New Look." *Journal of Financial and Quantitative Analysis* 14 (1979), pp. 119–35.

Milton, H. "Cost of Research Index, 1920–1965." *Operations Research* 14 (1966), pp. 977–91.

_____. "Cost of Research Index 1920–70." *Operations Research* 20 (1972), pp. 1–17.

Minasian, J. "Research and Development, Production Functions, and Rates of Return." *American Economic Review* 59 (May, 1969), pp. 80–85.

Modigliani, F., and M. Miller. "The Cost of Capital, Corporation Finance, and the Theory of Investment." *American Economic Review* 48 (1958), pp. 261–97.

Moody's Industrial Manual: 1977. New York: Moody's Investors Service, Inc., 1977.

Mueller, D. "The Firm Decision Process: An Econometric Investigation." *Quarterly Journal of Economics* 81 (1967), pp. 58–87.

Murphy, E. F. "The Interaction of Company and Federal Funds for the Performance of Research and Development in Large Manufacturing Companies." Ph.D. Dissertation, Duke University, 1965.

Myers, S. "Interactions of Corporate Financing and Investment Decisions—Implications for Capital Budgeting." *Journal of Finance* 29 (1974), pp. 1–25.

Nakamura, Alice, and Masao Nakamura. "On the Firm's Production, Capital Structure, and Demand for Debt." *Review of Economics and Statistics* 54 (1982), pp. 384–93.

Nelson, R. "The Simple Economics of Basic Scientific Research." *Journal of Political Economy,* 67 (1959), pp. 297–306.

_____, and S. Winter. "Forces Generating and Limiting Concentration Under Schumpeterian Competition." *Bell Journal of Economics* 9 (1978), pp. 524–48.

Organization for Economic Cooperation and Development. *Trends in Industrial R and D in Selected OECD Member Countries, 1967–75.* Paris: OECD, 1979.

Phillips, A. "A Critique of Empirical Studies of Relations Between Market Structure and Profitability." *Journal of Industrial Economics* 24 (1976), pp. 241–49.

Roberts, E. G. *The Dynamics of Research and Development.* New York: Harper and Row, 1964.

Robichek, A., J. McDonald, and R. Higgins. "Some Estimates of the Cost of Capital to the Electric Utility Industry, 1954–57: Comment." *American Economic Review* 57 (1967), pp. 1278–88.

Romeo, A., and W. McEachern. "Stockholder Control, Uncertainty, and the Allocation of Resources to Research and Development." *Journal of Industrial Economics* 26 (1978), pp. 349–61.

Scherer, F. M. *Industrial Market Structure and Economic Performance.* Chicago: Rand-McNally, 1980.

_____, and D. Ravenscraft. "The Lag Structure of Returns to R & D." Presented at the meeting of the American Economic Association, December 30, 1981.

Schmitt, R. W., and P. J. Stewart. "Unrealities in the Energy R & D Program." *Research Management* 21 (1978), pp. 7–9.

Schott, K. "The Relations Between Industrial Research and Development and Factor Demands." *Economic Journal* 88 (1978), pp. 85–106.

Schrieves, R. "Market Structure and Innovation: A New Perspective." *Journal of Industrial Economics* 26 (1978), pp. 329–46.

Shephard, R. *Theory of Cost and Production Functions.* Princeton: Princeton University Press, 1970.

Standard and Poor's Register of Corporations, 1977.

Survey of Current Business. *Business Statistics,* 1977 edition.

Switzer, L. "Price Indices for Industrial R & D: A Project Level Approach." *R & D Management* 13 (1983), pp. 101–6.

_____. "The Determinants of Industrial R & D: A Funds Flow Simultaneous Equation Approach." *Review of Economics and Statistics* 66 (1984), pp. 163–68.

Teece, D., and H. Armour. "Innovation and Divestiture in the U.S. Oil Industry." In D. Teece, ed., *R & D in Energy: Implications of Petroleum Industry Reorganization.* Palo Alto: Institute for Energy Studies, 1977, pp. 7–97.

Teem, J. M. "The ERDA—Its Role, Function, and Plans." *Research Management* 18 (1975), pp. 25–29.

Terleckyj, N. E. *Effects of R & D on the Productivity Growth of Industries: An Exploratory Study.* Washington, D.C.: National Planning Association, 1974.

_____. "Direct and Indirect Effects of Industrial Research and Development on the Productivity Growth of Industries." In Kendrick and Vaccara, eds., *New Developments in Productivity Measurement and Analysis.* Chicago: University of Chicago Press, 1980, pp. 359–77.

Theil, H. *Principles of Econometrics.* New York: John Wiley, 1971.

Tilton, J. "Research and Development in Industrial Growth: A Comment." *Journal of Political Economy* 81 (1973), pp. 1245–48.

Tobin, J. and W. Brainard. "Pitfalls in Financial Model Building." *American Economic Review* 58 (May, 1968), pp. 99–122.

U.S. Government, Department of Commerce. *1972 Census of Manufactures.* Washington, D.C.: U.S. Government Printing Office, 1976.

_____. *Code of Federal Regulations,* Title 41, Public Contracts and Property Management. Washington, D.C.: U.S. Government Printing Office, 1979.

_____. Economic Report of the President. *1972 Annual Report of the Council of Economic Advisers.* Washington, D.C.: U.S. Government Printing Office, 1972.

_____. Economic Report of the President. *1981 Annual Report of the Council of Economic Advisers.* Washington, D.C.: U.S. Government Printing Office, 1981.

_____. National Science Foundation. *Resources for Scientific Activities at Universities and Colleges.* Washington, D.C.: U.S. Government Printing Office, 1971.

_____. National Science Foundation. *Federal R & D Funding by Budget Function, Fiscal Years 1979-81.* Washington, D.C.: U.S. Government Printing Office, 1980.

_____. National Science Foundation. *Research and Development in Industry.* Washington, D.C.: U.S. Government Printing Office, various editions.

_____. *Research and Development Directory,* 1977 edition. Washington, D.C.: Government Data Publications, 1978.

Weston, J. "A Test of Capital Propositions." *Southern Economic Journal* 30 (1963), pp. 105–12.

Williamson, O. *Markets and Hierarchies: Analysis and Antitrust Implications.* New York: The Free Press, 1975.

_____. "Transaction Cost Economics: The Governance of Contractual Relations." *Journal of Law and Economics* 22 (1974), pp. 233–61.

_____. "The Economics of Organization: The Transaction Cost Approach." Centre for the Study of Organizational Innovation, University of Pennsylvania, Discussion Paper #82, August, 1980.

Wilson, R. "The Effect of Technological Environment and Product Rivalry on R & D Effort and Licensing of Inventions." *Review of Economics and Statistics* 49 (1977), pp. 171–78.

Wood, R. "Research and Development Expenditures Under the Economic Recovery Tax Act." *TAXES—The Tax Magazine* 59 (1981), pp. 777–83.

Index